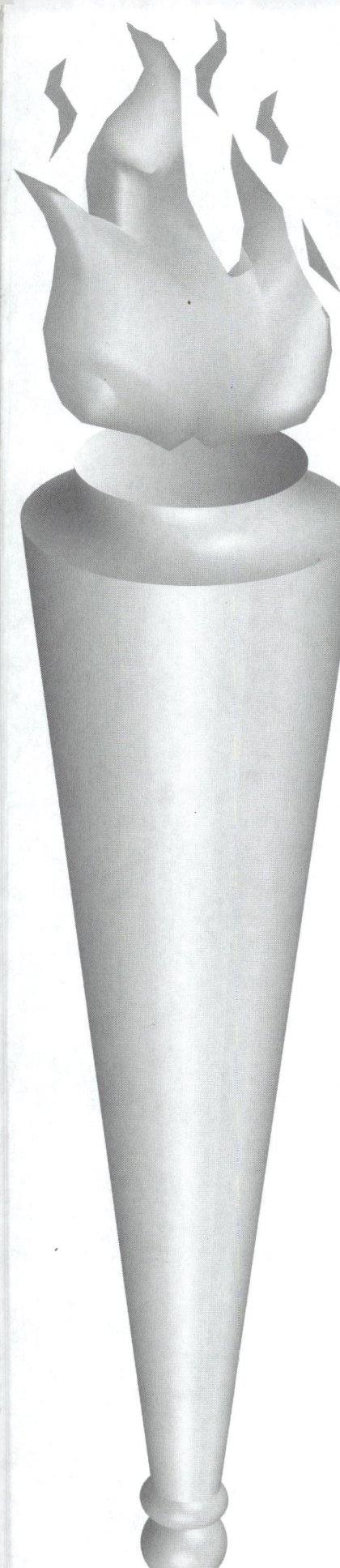

Be prepared...
To learn...
To succeed...

Get **REA**dy. It all starts here. REA's preparation for the TAKS is **fully aligned** with the Texas TEKS guidelines adopted by the Texas Education Agency.

Visit us online at
www.rea.com

Ready, Set, Go!™

TAKS
Mathematics, Exit Level
2nd Edition

Staff of Research & Education Association

Research & Education Association

For all references in this book, Texas Assessment of Knowledge and Skills™ and TAKS™ are trademarks of the Texas Education Agency (TEA). The test objectives presented in this book were created and implemented by the Student Assessment Division of the TEA. For further information, visit the TEA website at *www.tea.state.tx.us*.

Research & Education Association
61 Ethel Road West
Piscataway, New Jersey 08854
E-mail: info@rea.com

Ready, Set, Go!
Texas TAKS™ Mathematics Test, Exit Level

Copyright © 2009 by Research & Education Association, Inc.

Prior edition copyright © 2008. All rights reserved. No part of this book may be reproduced in any form without permission of the publisher.

Printed in the United States of America

Library of Congress Control Number 2008925157

ISBN 13: 978-0-7386-0444-2
ISBN 10: 0-7386-0444-5

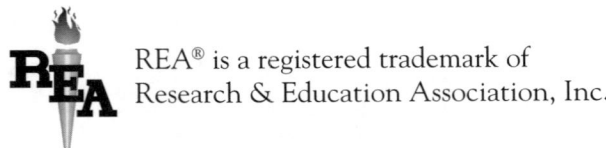

REA® is a registered trademark of Research & Education Association, Inc.

G08

About Research & Education Association

Founded in 1959, Research & Education Association is dedicated to publishing the finest and most effective educational materials—including software, study guides, and test preps—for students in middle school, high school, college, graduate school, and beyond. Today, REA's wide-ranging catalog is a leading resource for teachers, students, and professionals. We invite you to visit us at *www.rea.com* to find out how REA is making the world smarter.

Acknowledgments

We would like to thank REA's Larry B. Kling, Vice President, Editorial, for supervising development; Pam Weston, Vice President, Publishing, for setting the quality standards for production integrity and managing the publication to completion; Michael Reynolds, Managing Editor, for project management and preflight editorial review; Christine Saul, Senior Graphic Artist, for cover design; Jeff LoBalbo, Senior Graphic Artist, for post-production file mapping; and Caragraphics for typesetting this edition.

Contents

Introduction: Passing the TAKS Mathematics Test 1
About This Book 1
About the Test 1
How To Use This Book 3
Standards 4

Chapter 1 Numbers and Operations 13
    Types of Numbers 13
    Absolute Value 14
    Equivalent Numbers 14
    Scientific Notation 20
    Square Roots and Radicals 22
Let's Review 1: Numbers and Operations 23
    Order of Operations 25
    Analyzing Numbers 26
Chapter 1 Practice Problems 27
Chapter 1 Answer Explanations 31

Chapter 2 Equations and Inequalities 35
    Objectives 35
    Order of Operations 36
    Properties 37
    Associative Property 38
    Expressions 40
    Equations 42
    Inequalities 44
    Patterns 45
Let's Review 2: Equations 46
Chapter 2 Practice Problems 50
Chapter 2 Answer Explanations 53

Chapter 3 Functions, Part 1 57
- Objectives 57
- Dependent and Independent Variables 58
- Functional Relationships 59
Let's Review 3: Functions 62
- Graphing Inequalities 65
- Selecting Expressions and Equations 66
- Linear Equations 67
Let's Review 4: Linear Equations 69
Chapter 3 Practice Problems 73
Chapter 3 Answer Explanations 78

Chapter 4 Functions, Part 2 81
- Objectives 81
- Linear Functions 83
- Quadratic Functions 87
- Parent Functions 88
- Slope 94
- Slope-Intercept Form of a Linear Function 94
Let's Review 5: Slope 96
- Scatterplots 98
Let's Review 6: Graphs and Equations 100
Chapter 4 Practice Problems 105
Chapter 4 Answer Explanations 109

Chapter 5 Geometry and Spatial Sense, Part 1 111
- Objectives 111
- Plane Figures 112
- Congruent Figures 114
- Similar Figures 114
Let's Review 7: Plane Figures 116
- Lines 118
- Angles 119
- Angle Relationships 120
Let's Review 8: Lines and Angles 122
- Triangles 123
- Pythagorean Theorem 124
- Circles 126
Let's Review 9: Triangles and Circles 127
Chapter 5 Practice Problems 129
Chapter 5 Answer Explanations 131

Contents

Chapter 6 Geometry and Spatial Sense, Part 2 133
Objectives 133
Three-Dimensional Figures 134
The Coordinate Plane 136
Transformations 137
Let's Review 10: Three-Dimensional Figures 140
Nets 142
Alternate Views 142
Chapter 6 Practice Problems 144
Chapter 6 Answer Explanations 149

Chapter 7 Measurement 151
Objectives 151
Perimeter 152
Circumference 153
Let's Review 11: Perimeter 154
Area 156
Volume 158
Surface Area 160
Let's Review 12: Area 161
Chapter 7 Practice Problems 163
Chapter 7 Answer Explanations 166

Chapter 8 Data Analysis and Probability 169
Objectives 169
Probability 170
Let's Review 13: Probability 173
Central Tendency: Mean, Median, and Mode 176
Range 177
Let's Review 14: Central Tendency 179
Charts and Graphs 181
Venn Diagrams 183
Let's Review 15: Graphs 185
Real-Life Situations 187
Let's Review 16: Discounts and Sale Prices 190
Chapter 8 Practice Problems 193
Chapter 8 Answer Explanations 196

Chapter 9 Problem Solving **201**
    Objectives 201
Let's Review 17: Problem Solving 203
Chapter 9 Practice Problems 208
Chapter 9 Answer Explanations 212

Practice Test 1 .... **217**
Practice Test 1 Reference Sheets **218**
Practice Test 1 Answer Document **243**
Practice Test 1 Answer Explanations **245**
Practice Test 2 .... **253**
Practice Test 2 Reference Sheets **254**
Practice Test 2 Answer Document **275**
Practice Test 2 Answer Explanations **279**
Index **292**

Introduction

Passing the TAKS Mathematics Test

About This Book

This book will provide you with an accurate and complete representation of the Texas Assessment of Knowledge and Skills (TAKS) in mathematics. Inside you will find reviews that are designed to provide you with the information and strategies needed to do well on these tests. Two practice tests based on the official TAKS are included. These practice tests contain every type of question that you can expect to encounter on the TAKS. Following each practice test, you will find answer keys with detailed explanations designed to help you completely understand the material. Answer keys and explanations at the end of each chapter will help you to understand the exercises within the chapter.

About the Test

Who Takes These Tests and What Are They Used For?

The Texas Assessment of Knowledge and Skills is given in grades 3 through 11. The TAKS is aligned to the Texas Essential Knowledge and Skills (TEKS) curriculum framework objectives that were adopted by the State Board of Education to test students' knowledge in mathematics, reading/English language arts, science, and social studies. To earn a Texas high school diploma, students must demonstrate that they have achieved the standards on the Exit Level tests.

Is There a Registration Fee?

No. Because all Texas public high school students are required to take the TAKS and pass each part of the test to receive a high-school diploma, no fee is required.

When and Where Is the Test Given?

The TAKS is administered annually in selected grades. Testing periods are scheduled for the spring of each year, and retests are administered in the fall. All students must pass the grade 11 tests to graduate. Students who do not pass these tests will be retested in grade 12. Tests are given in the individual schools. The TAKS is not timed, but students are expected to finish the test within a couple of hours.

Test Accommodations and Special Situations

All students educated with public funds are required to take the TAKS. Very rarely is an exemption granted. Special education and disabled students are required to take the TAKS. Every effort is made to accommodate students with disabilities who take the TAKS and seek a standard high-school diploma. For example, the timing or scheduling of the test can be altered according to a student's medical needs. Changes also can be made to the test setting, how the test is presented, and the manner in which the student responds to test questions to accommodate specific needs. Students with significant disabilities who are unable to take the standard TAKS tests even with accommodations participate in the TAKS Alternate Assessment (TAKS-Alt). English language learners (ELL) are also required to take the TAKS. Spanish versions of the TAKS are available in grades 3 through 6. After grade 6, the TAKS is available only in English.

Additional Information and Support

Additional resources to help you prepare to take the TAKS can be found on the Texas Department of Education website at *http://www.tea.state.tx.us/*

How to Use This Book

What Do I Study First?

Read over the review sections and the suggestions for test taking. Studying the review sections thoroughly will reinforce the basic skills you need to do well on the test. Be sure to take the practice tests to become familiar with the format and procedures involved with taking the actual TAKS.

When Should I Start Studying?

It is never too early to start studying for the TAKS. The earlier you begin, the more time you will have to sharpen your skills. Do not procrastinate! Cramming is not an effective way to study because it doesn't allow you the time needed to learn the test material. The sooner you learn the format of the exam, the more time you will have to familiarize yourself with the exam content.

Overview of the TAKS in Mathematics

Each TAKS test has 60 questions based on the learning standards in the Texas Essential Knowledge and Skills for mathematics. The TAKS objectives for mathematics identify the four major content strands listed below.

- Number Sense and Quantitative Reasoning
- Algebra
- Geometry
- Probability and Statistics

Most test items on the TAKS test are multiple-choice format with four answer choices. There are a limited number of open-response questions for which students fill in responses in a grid.

Standards

Objective 1: The student will describe functional relationships in a variety of ways.

A(b)(1) Foundations for functions. The student understands that a function represents a dependence of one quantity on another and can be described in a variety of ways.

(A) The student describes independent and dependent quantities in functional relationships.

(B) The student [gathers and records data, or] uses data sets, to determine functional (systematic) relationships between quantities.

(C) The student describes functional relationships for given problem situations and writes equations or inequalities to answer questions arising from the situations.

(D) The student represents relationships among quantities using [concrete] models, tables, graphs, diagrams, verbal descriptions, equations, and inequalities.

(E) The student interprets and makes inferences from functional relationships.

Objective 2: The student will demonstrate an understanding of the properties and attributes of functions.

A(b)(2) Foundations for functions. The student uses the properties and attributes of functions.

(A) The student identifies [and sketches] the general forms of linear ($y = x$) and quadratic ($y = x^2$) parent functions.

(B) For a variety of situations, the student identifies the mathematical domains and ranges and determines reasonable domain and range values for given situations.

(C) The student interprets situations in terms of given graphs [or creates situations that fit given graphs].

(D) In solving problems, the student [collects and] organizes data, [makes and] interprets scatter plots, and models, predicts, and makes decisions and critical judgments.

A(b)(3) Foundations for functions. The student understands how algebra can be used to express generalizations and recognizes and uses the power of symbols to represent situations.

(A) The student uses symbols to represent unknowns and variables.

(B) Given situations, the student looks for patterns and represents generalizations algebraically.

A(b)(4) Foundations for functions. The student understands the importance of the skills required to manipulate symbols in order to solve problems and uses the necessary algebraic skills required to simplify algebraic expressions and solve equations and inequalities in problem situations.

(A) The student finds specific function values, simplifies polynomial expressions, transforms and solves equations, and factors as necessary in problem situations.

(B) The student uses the commutative, associative, and distributive properties to simplify algebraic expressions.

Objective 3: The student will demonstrate an understanding of linear functions.

A(c)(1) Linear functions. The student understands that linear functions can be represented in different ways and translates among their various representations.

(A) The student determines whether or not given situations can be represented by linear functions.

(B) The student translates among and uses algebraic tabular, graphical, or verbal descriptions of linear function.

A(c)(2) Linear functions. The student understands the meaning of the slope and intercepts of linear functions and interprets and describes the effects of changes in parameters of linear functions in real-world and mathematical situations.

(A) The student develops the concepts of slope as a rate of change and determines slopes from graphs, tables, and algebraic expressions.

(B) The student interprets the meaning of slope and intercepts in situations using data, symbolic representations, or graphs.

(C) The student investigates, describes, and predicts the effects of changes in *m* and *b* on the graph of $y = mx + b$.

(D) The student graphs and writes equations of lines given characteristics, such as two points, a point and a slope, or a slope and *y*-intercept.

(E) The student determines the intercepts of linear functions from graphs, tables, and algebraic representations.

(F) The student interprets and predicts the effects of changing slope and *y*-intercept in applied situations.

(G) The student relates direct variation to linear functions and solves problems involving proportional change.

Objective 4: The student will formulate and use linear equations and inequalities.

A(c)(3) Linear functions. The student formulates equations and inequalities based on linear functions, uses a variety of methods to solve them, and analyzes the solutions in terms of the situation.

(A) The student analyzes situations involving linear functions and formulates linear equations or inequalities to solve problems.

(B) The student investigates methods for solving linear equations and inequalities using [concrete] models, graphs, and the properties of equality, selects a method, and solves the equations and inequalities.

(C) For given contexts, the student interprets and determines the reasonableness of solutions to linear equations and inequalities.

A(c)(4) **Linear functions.** The student formulates systems of linear equations from problem situations, uses a variety of methods to solve them, and analyzes the solutions in terms of the situation.

(A) The student analyzes situations and formulates systems of linear equations to solve problems.

(B) The student solves systems of linear equations using [concrete] models, graphs, tables, and algebraic methods.

(C) For given contexts, the student interprets and determines the reasonableness of solutions to systems of linear equations.

Objective 5: The student will demonstrate an understanding of quadratic and other nonlinear functions.

A(d)(1) **Quadratic and other nonlinear functions.** The student understands that the graphs of quadratic functions are affected by the parameters of the function and can interpret and describe the effects of changes in the parameters of quadratic functions.

(B) The student investigates, describes, and predicts the effects of changes in a on the graph of $y = ax^2$.

(C) The student investigates, describes, and predicts the effects of changes in c on the graph of $y = x^2 + c$.

(D) For problem situations, the student analyzes graphs of quadratic functions and draws conclusions.

A(d)(2)   Quadratic and other nonlinear functions. The student understands there is more than one way to solve a quadratic equation and solves them using appropriate methods.

   (A)   The student solves quadratic equations using [concrete] models, tables, graphs, and algebraic methods.

   (B)   The student relates the solutions of quadratic equations to the roots of their functions.

A(d)(3)   Quadratic and other nonlinear functions. The student understands there are situations modeled by functions that are neither linear nor quadratic and models the situations.

   (A)   The student uses [patterns to generate] the laws of exponents and applies them in problem-solving situations.

Objective 6: The student will demonstrate an understanding of geometric relationships and spatial reasoning.

   G(b)(4)   Geometric structures. The student uses a variety of representations to describe geometric relationships and solve problems.

   (A)   The student selects an appropriate representation ([concrete], pictorial, graphical, verbal, or symbolic) in order to solve problems.

   G(c)(1)   Geometric patterns. The student identifies, analyzes, and describes patterns that emerge from two- and three-dimensional geometric figures.

   (A)   The student uses numeric and geometric patterns to make generalizations about geometric properties, including properties of polygons, ratios in similar figures and solids, and angle relationships in polygons and circles.

   (B)   The student uses the properties of transformations and their compositions to make connections between mathematics and the real world in applications such as tessellations or fractals.

(C) The student identifies and applies patterns from right triangles to solve problems, including special right triangles (45-45-90 and 30-60-90) and triangles whose sides are Pythagorean triples.

G(e)(3) **Congruence and the geometry of size**. The student applies the concept of congruence to justify properties of figures and solve problems.

(A) The student uses congruence transformations to make conjectures and justify properties of geometric figures.

Objective 7: The student will demonstrate an understanding of two- and three-dimensional representations of geometric relationships and shapes.

G(d)(1) **Dimensionality and the geometry of location**. The student analyzes the relationship between three-dimensional objects and related two-dimensional representations and uses these representations to solve problems.

(B) The student uses nets to represent [and construct] three-dimensional objects.

(C) The student uses top, front, side, and corner views of three-dimensional objects to create accurate and complete representations and solve problems.

G(d)(2) **Dimensionality and the geometry of location**. The student understands that coordinate systems provide convenient and efficient ways of representing geometric figures and uses them accordingly.

(A) The student uses one- and two-dimensional coordinate systems to represent points, lines, line segments, and figures.

(B) The student uses slopes and equations of lines to investigate geometric relationships, including parallel lines, perpendicular lines, and [special segments of] triangles and other polygons.

(C) The student [develops and] uses formulas, including distance and midpoint.

G(e)(2) **Congruence and the geometry of size.** The student analyzes properties and describes relationships in geometric figures.

(D) The student analyzes the characteristics of three-dimensional figures and their component parts.

Objective 8: The student will demonstrate an understanding of the concepts and uses of measurement and similarity.

G(e)(1) **Congruence and the geometry of size.** The student extends measurement concepts to find area, perimeter, and volume in problem situations.

(A) The student finds areas of regular polygons and composite figures.

(B) The student finds areas of sectors and arc lengths of circles using proportional reasoning.

(C) The student [develops, extends and] uses the Pythagorean Theorem.

(D) The student finds surface area and volumes of prisms, pyramids, spheres, cones, and cylinders in problem situations.

G(f)(1) **Similarity and the geometry of shape.** The student applies the concepts of similarity to justify properties of figures and solve problems.

(A) The student uses similarity properties and transformations to [explore and] justify conjectures about geometric figures.

(B) The student uses ratios to solve problems involving similar figures.

(C) In a variety of ways, the student [develops], applies, and justifies triangle similarity relationships, such as right triangle ratios, [trigonometric ratios], and Pythagorean triples.

(D) The student describes the effect on perimeter, area, and volume when length, width, or height of a three-dimensional solid is changed and applies this idea in solving problems.

Objective 9: The student will demonstrate an understanding of percents, proportional relationships, probability, and statistics in application problems.

(8.3) **Patterns, relationships, and algebraic thinking.** The student identifies proportional relationships in problem situations and solves problems. The student is expected to

(B) estimate and find solutions to application problems involving percents and proportional relationships, such as similarity and rates.

(8.11) **Probability and statistics.** The student applies the concepts of theoretical and experimental probability to make predictions. The student is expected to

(A) find the probabilities of compound events (dependent and independent); and

(B) use theoretical probabilities and experimental results to make predictions and decisions.

(8.12) **Probability and statistics.** The student uses statistical procedures to describe data. The student is expected to

(A) select the appropriate measure of central tendency to describe a set of data for a particular purpose; and

(C) construct circle graphs, bar graphs, and histograms, with and without technology.

(8.13) **Probability and statistics.** The student evaluates predictions and conclusions based on statistical data. The student is expected to

(B) recognize misuses of graphical or numerical information and evaluate predictions and conclusions based on data analysis.

Objective 10: The student will demonstrate an understanding of the mathematical processes and tools used in problem solving.

(8.14) **Underlying processes and mathematical tools**. The student applies Grade 8 mathematics to solve problems connected to everyday experiences, investigations into other disciplines, and activities in and outside of school. The student is expected to

(A) identify and apply mathematics to everyday experiences, to activities in and outside of school, with other disciplines, and with other mathematical topics;

(B) use a problem-solving model that incorporates understanding the problem, making a plan, carrying out the plan, and evaluating the solution for reasonableness; and

(C) select or develop an appropriate problem-solving strategy from a variety of different types, including drawing a picture, looking for a pattern, systematic guessing and checking, acting it out, making a table, working a simpler problem, or working backwards to solve a problem.

(8.15) **Underlying processes and mathematical tools**. The student communicates about Grade 8 mathematics through informal and mathematical language, representations, and models. The student is expected to

(A) communicate mathematical ideas using language, efficient tools, appropriate units, and graphical, numerical, physical, or algebraic mathematical models.

(8.16) **Underlying processes and mathematical tools**. The student uses logical reasoning to make conjectures and verify conclusions. The student is expected to

(A) make conjectures from patterns or sets of examples and nonexamples; and

(B) validate his/her conclusions using mathematical properties and relationships.

Chapter 1
Numbers and Operations

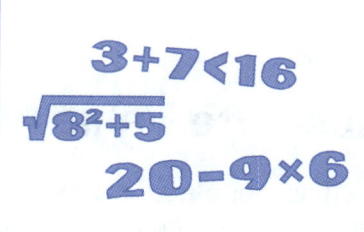

Questions on the Texas Assessment of Knowledge and Skills (TAKS) will be about numbers in different forms, such as fractions and percents. You may be asked to determine the percent of a discount, for example, or to determine what percentage of students surveyed enjoy a certain kind of music. Some questions will involve numbers raised to a power or numbers written in scientific notation. This chapter will give you a review of numbers in different forms.

Types of Numbers

Integers are whole numbers and their opposites. A positive number's opposite is its negative. The numbers below are pairs of opposites:

1	−1
2	−2
3	−3
4	−4
5	−5

Real numbers are numbers that can be placed on a number line. Real numbers are grouped into two categories: rational and irrational. **Rational numbers** include whole numbers, fractions and decimals, even if these numbers are repeating, meaning they don't terminate but continuously repeat the same pattern. Every rational number can be written as a ratio of two integers. **Irrational numbers** are decimals that don't repeat or terminate in a logical manner. For example, the following numbers are irrational:

$$2.3459564646332\ldots$$
$$0.999088432\ldots$$
$$\pi = 3.14159\ldots$$

13

Absolute Value

The **absolute value** of a number is the number of units it is from zero on the number line. Therefore, to find a number's absolute value, it helps to imagine a number line, such as the one shown below. You can see that the absolute value of -2 is 2 because -2 is 2 units from 0.

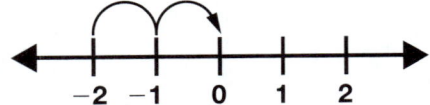

Equivalent Numbers

Numbers that are **equivalent** have the same value. With some numbers, it is easy to see that they are equivalent. For example, you know that $7 = 7$, and you probably know that $\frac{5}{5}$ is equivalent to 1.

Determining whether numbers are equivalent when they are in different forms is more difficult, however. You might not know right away that 8^3 is equivalent to 512.

The best way to determine whether numbers are equivalent is put them into the same form. The following are some guidelines to help you do this.

Fractions

A fraction has a "top" number and a "bottom" number. The **numerator** (top number) of a fraction tells how many parts of the whole you have. The **denominator** (bottom number) tells into how many parts the whole is evenly divided. For example, the fraction $\frac{2}{3}$ tells you that you have 2 out of 3 equal parts.

If the denominators of two fractions are the same, the fraction with the *larger* numerator is the larger fraction. For example, $\frac{3}{7}$ is larger than $\frac{2}{7}$.

Equivalent fractions are fractions that represent the same part of the whole. For example, $\frac{1}{2}$ is equivalent to $\frac{2}{4}$ and to $\frac{3}{6}$.

If two fractions have different numerators and denominators, you can determine whether they are equivalent by making the denominators the same. To do this, you

Chapter 1: Numbers and Operations

must realize that 1 times any number equals that same number. Multiply one of the fractions by the equivalent of 1 so that the denominators of the two fractions are the same. Then compare the results to see whether the fractions are equivalent.

To multiply two fractions, multiply the two numerators to get the new numerator, and then multiply the two denominators to get the new denominator, so $\frac{2}{5} \times \frac{3}{4} = \frac{6}{20}$.

For example, to determine whether $\frac{2}{3}$ and $\frac{4}{6}$ are equivalent, multiply $\frac{2}{3}$ by $\frac{2}{2}$ so it has the same denominator (6) as $\frac{4}{6}$. (Note that $\frac{2}{2}$ is equivalent to 1, so it doesn't change the value of $\frac{2}{3}$.)

$$\frac{2}{3} \times \frac{2}{2} = \frac{4}{6}$$

Therefore, $\frac{2}{3} = \frac{4}{6}$, and the two fractions are equivalent.

Let's try another example: Determine whether $\frac{2}{3}$ is equivalent to $\frac{5}{9}$.

Multiply $\frac{2}{3}$ by $\frac{3}{3}$ so it has the same denominator as $\frac{5}{9}$.

$$\frac{2}{3} \times \frac{3}{3} = \frac{6}{9}$$

$\frac{6}{9}$ is larger than $\frac{5}{9}$, so the fractions are not equivalent.

Sometimes you may have to multiply each fraction by a different equivalent of 1 for them both to have the same denominators so you can compare them. For example, to determine whether $\frac{2}{3}$ and $\frac{4}{5}$ are equivalent fractions, do the following multiplications:

$$\frac{2}{3} \times \frac{5}{5} = \frac{10}{15}$$
$$\frac{4}{5} \times \frac{3}{3} = \frac{12}{15}$$

Thus, you can see that $\frac{2}{3}$ is not equivalent to $\frac{4}{5}$; in fact, it is smaller.

If you're asked to compare two or more mixed numbers (a mixed number has a whole number and a fraction, such as $1\frac{1}{2}$), the one with the larger whole number is the greater number. For example:

$$2\frac{1}{3} \text{ is greater than } 1\frac{1}{3}$$

If the whole number parts are the same, use the method you just learned to compare the fractional parts to determine which mixed number is larger.

Decimals

A mixed decimal number, such as 3.14, includes a decimal point and has two parts. The part to the left of the decimal point is a whole number, and the part to the right of the decimal point is called a decimal. A **decimal** is not a whole number, but is a portion of a whole number, and has a value less than 1. Therefore, the number 3 is greater than the number .33. The decimal .33 can also be written as 0.33, indicating that there is no whole number with the decimal.

The decimal system is based on the number 10 (this probably has to do with the fact that most humans have 10 fingers). Each digit in a decimal number has a value assigned to its "place." To the left of the decimal point, the digits appear as you are used to seeing them (ones, tens, hundreds, etc.), but to the right of the decimal point they are fractions, so they are tenths, hundredths, thousandths, etc.

So for the decimal number 4.25, the 4 is a whole number, the 2 is tenths, and the 5 is hundredths. You could put these last two together and say "twenty-five hundredths." You would read 4.25 as "four point two five," or "four and twenty-five hundredths."

Do you know which is greater, .334 or .3? To determine which decimal is greater, align the decimal points vertically, like this:

$$.334$$
$$.3$$

Then fill in the empty place values with zeros so both numbers have the same number of digits before you do the comparison:

.334
.300

Which decimal is greater? If you said .334, you're correct! This method is similar to comparing whole numbers, but you must remember to add the zeros to the ends of the decimal fractions so each decimal has the same number of digits.

To compare two mixed decimals, the decimal with the greater whole number is always larger. For example, 2.334 is greater than 1.945. If the two mixed decimals have the same whole number, use the method you just learned to compare the decimals to determine which is greater.

For example, to compare 1.4 and 1.36, align the decimal points of the numbers vertically and fill in zeros if necessary:

1.40
1.36

Now you can see that 1.4 is definitely greater.

On the TAKS, you may be asked to compare numbers in different forms, such as fractions and decimals. When comparing a fraction and a decimal, convert the fraction into a decimal by dividing the denominator into the numerator. Try converting the fractions $\frac{3}{4}$ and $\frac{5}{6}$ into decimals:

If you got . . .

$$\frac{3}{4} = .75$$

$$\frac{5}{6} = .8333333$$

. . . you are right.

Terminating decimals are decimals that stop. For example, .75 is a terminating decimal. **Repeating decimals** keep on going in a pattern. .8333333 is a repeating decimal (the 3 keeps repeating). The line over the last number (or group of numbers) indicates that a decimal is a repeating decimal, and the numbers under the line are the repeating numbers. Thus, $.833333\overline{3} = .8\overline{33} = .8\overline{3}$

Percents

A **percent** has a percent sign (%) and refers to a portion out of one hundred. For example, 75% means 75 out of 100. Determining which of two (or more) percents is greater is sometimes easy. For example, 75% is obviously greater than 65%.

As you learned earlier in this chapter, to find an equivalent number or to compare numbers in different forms, convert the numbers to the same form. Usually, it is easiest to convert to decimals to compare numbers of different forms.

To convert a percentage into a decimal, move the decimal point to the left two places. (These places represent the two zeros in 100.) Look at the following examples:

$$32\% = .32$$
$$75\% = .75$$
$$210\% = 2.10$$

Fill in zeros if necessary:

$$5\% = .05$$
$$.3\% = .003$$

Now, suppose you need to convert a decimal to a percentage. You would move the decimal point two places to the *right*. Don't forget to write the percent sign.

$$.20 = 20\%$$

If you need to convert a percentage to a fraction, put the percentage over 100. Then **reduce the fraction**, if possible, by the following method: Think of a number that divides evenly into both the numerator and denominator (this number is called a common factor). Do that division, and the result gives you the reduced fraction.

So for 20%, the calculation would be: $20\% = \frac{20}{100}$. Then divide both numerator and denominator by 20 to get $\frac{1}{5}$. So $20\% = \frac{1}{5}$.

Let's say that you didn't recognize right off that 20 divides into both 20 and 100 in the example above. Let's say you thought of 10 instead. Then the calculation would be: $20\% = \frac{20}{100} = \frac{2}{10}$. Perhaps now you see that 2 will divide into both the 2 and 10. The result will be $\frac{2}{10} = \frac{1}{5}$, the same result as when you reduced the fraction by using 20 as the common factor: $\frac{20}{100} = \frac{2}{10} = \frac{1}{5}$.

Powers

The power (indicated by a raised number, called an **exponent**) tells you how many times the number appears when it is multiplied by itself. Thus, the first power of any number is equal to itself. For example,

$$8^1 = 8$$

When you raise a number to the second power, you **square** the number. When you square a number, you multiply it by itself, as in this example:

$$8^2 = 8 \times 8$$

Note that the 8 appears two times when it is squared.

When you raise a number to the third power, you **cube** the number. To cube a number, multiply it by itself and then by itself again:

$$8^3 = 8 \times 8 \times 8$$

Note here that the 8 appears three times when it is cubed.

You can keep raising numbers to higher and higher powers, as in this example:

$8^9 = 8 \times 8 \times 8 \times 8 \times 8 \times 8 \times 8 \times 8 \times 8$, where the number 8 appears nine times.

Let's try to multiply $8^3 \times 8^2$. You get $(8 \times 8 \times 8) \times (8 \times 8)$, which is 8^5. So you can see that if you multiply two like numbers with different powers, you can simply add the powers: $8^3 \times 8^2 = 8^{2+3} = 8^5$. Likewise, when you divide two like numbers with different powers, you subtract the powers: $8^5 \div 8^2 = 8^{5-2} = 8^3$.

Scientific Notation

Scientific notation is a type of shorthand for writing numbers that are very large or very small. These numbers usually contain many zeros. The table below shows the exponents of 10 used in scientific notation. (Note that for positive powers of 10, the exponent tells you how many zeros follow the "1" in the number value.)

Number Value	Power of 10
1	10^0
10	10^1
100	10^2
1,000	10^3
10,000	10^4
100,000	10^5
1,000,000	10^6

But what about a number like 2,500,000? How would you write this number in scientific notation? Move the decimal point from the end to the *left* until there is just one digit to the left of it. Here we would move the decimal point until it is between the 2 and the 5. The number of places you move the decimal point (in this case, 6) is the exponent of 10.

$$2{,}500{,}000 = 2.5 \times 10^6$$

Let's try another number. How would you write 32,000 in scientific notation? Remember to move the decimal point until it is between the 3 and the 2. The number of places you moved it is the exponent of 10.

$$32{,}000 = 3.2 \times 10^4$$

Now, let's go the other way, and write out the number 4.8×10^5.

This time you need to move the decimal point to the *right* five places. Remember to fill in zeros.

$$4.8 \times 10^5 = 480{,}000$$

Very small numbers (decimals) can also be represented with scientific notation, but in a slightly different way. A negative exponent is used with the number 10 to indicate a decimal.

The following table shows how scientific notation is used to represent very small numbers. (Note that for negative powers of 10, the exponent tells you what "place" the 1 is in, counting to the right from the decimal point.)

Decimal	Power
0.1	10^{-1}
0.01	10^{-2}
0.001	10^{-3}
0.0001	10^{-4}
0.00001	10^{-5}
0.000001	10^{-6}

How do you think you would write 0.0062 using scientific notation? This time, you would move the decimal point to the *right* until it has only one non-zero digit before it. Here you would move the decimal point until it is between the 6 and the 2. You would need to move it three places, so your exponent would be −3.

$$0.0062 = 6.2 \times 10^{-3}$$

Similarly, to determine the decimal for 4.7×10^{-4}, move the decimal point to the left four places, filling in zeros as necessary:

$$4.7 \times 10^{-4} = .00047$$

Square Roots and Radicals

The **square root** of a number is the **inverse operation** of squaring a number (multiplying the number by itself). For example,

$$12^2 = 144, \text{ so } \sqrt{144} = 12$$

Not every number is a perfect square. This means you might not always get a whole number when you find the square root. For example,

$$\sqrt{3} = 1.7320508\cdots$$

Remember that the three dots mean the decimal doesn't end after the 8, but continues forever without repeating.

Radicals are symbols for roots. When adding or subtracting radicals, keep the number under the radical, called the **radicand**, the same and add or subtract the coefficients, the numbers in front of the square root sign. For example,

$$2\sqrt{3} + 6\sqrt{3} = 8\sqrt{3}$$

When multiplying or dividing radicals, follow the same rules you would when multiplying or dividing whole numbers. For example,

$$\sqrt{3} \times \sqrt{3} = \sqrt{9}$$

$$\sqrt{8} \times \sqrt{2} = \sqrt{16}$$

$$2\sqrt{2} \times 4\sqrt{5} = 8\sqrt{10}$$

Chapter 1: Numbers and Operations

Let's Review 1: Numbers and Operations

Complete each of the following questions. Use the Tip below each question to help you choose the correct answer. When you finish, check your answers with those at the end of Chapter 1.

1 The population of Texas is about 21 million.

What is 21 million expressed in scientific notation?

A 2.1×10^8
B 2.1×10^7
C 2.1×10^6
D 2.1×10^5

21,000,000
2.1 × 10⁶

TIP: Remember that with scientific notation, count the number of places you must move the decimal to get to 2.1. The number of places you move the decimal is the exponent of 10.

2 Which is another way to express 144?

F 12^2
G 4^4
H $\dfrac{1}{144}$
J 8^3

12 × 12 = 144

TIP: If you're not sure of the correct answer, begin by eliminating those that you know are incorrect.

4 × 4 × 4 × 4
16 × 4 = 64
64 × 4 = 256

8 × 8 × 8 =
64 × 8 = 512

24 TAKS   **Mathematics**

3 Which fraction is equal to .45?

A $\frac{1}{45}$

B $\frac{9}{20}$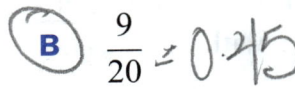

C $\frac{1}{4}$

D $\frac{1}{2} = 0.5$

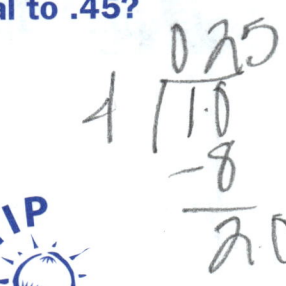

TIP Either convert the decimal to a fraction and reduce the fraction and compare the result to the answer choices, or use your calculator to find the decimal equivalent of each fraction in the answer choices by dividing the denominators into the numerators and compare the decimal results to .45.

(B is circled)

4 What is .25 expressed as a percentage?
Be sure to use the correct place value.

·25·

TIP To convert a decimal to a percentage, move the decimal point two places to the right.

Chapter 1: Numbers and Operations   25

5 The side of a triangle is √32 inches. Which point is closest to √32 on the number line?

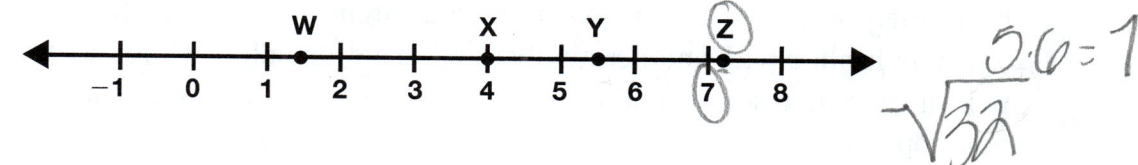

A W

B X

C Y

D Z

TIP

Find the square root of 32. Then choose the point that is closest to this number.

Order of Operations

Many questions involve more than one operation. To answer such questions correctly, you have to perform operations in the correct order. Look at this problem:

$$2 + (8 \times 3) = ?$$

If you solve the problem by doing the operations from left to right, you would get 30 because 2 + 8 is 10, and 10 × 3 = 30.

However, if you multiply 8 × 3 first and then add 2, the answer is 26. This example uses the correct order of operations as listed below.

Perform operations in this order:

1. Perform operations inside parentheses first. Within the parentheses, use the order in the steps below.

2. Next, perform powers from left to right.

3. Then multiply or divide from left to right.

4. Last, add or subtract from left to right.

Analyzing Numbers

For some problems on the TAKS, you will have to identify the results of operations performed on numbers. For example, you might be asked to choose a number that, when multiplied with another number, will have a result of 1. Some of these problems will involve inverse operations. An **inverse operation** is any operation that "undoes" other operation.

Addition and subtraction are inverse operations because adding and subtracting the same number results in no change. For example, $8 + 3 - 3 = 8$ because adding 3 and subtracting 3 are inverse operations that cancel each other.

Multiplying and dividing by the same non-zero number also results in no change, so they are inverse operations.

Chapter 1 Practice Problems

Complete each of the following practice problems. Check your answers at the end of this chapter. Be sure to read the answer explanations!

1 Which value is the greatest?

A. 7^3 — $7 \times 7 \times 7 = 343$

B. 7^4 — $7 \times 7 \times 7 \times 7 = 2401$

C. 8^6 — $8 \times 8 \times 8 \times 8 \times 8 \times 8 = 262,144$

D. 9^5 — $9 \times 9 \times 9 \times 9 \times 9 = 59,049$

(C is circled)

2 Which answer is .64 written as a percent?

F. .64%

G. 6.4% = 0.0064 $\frac{100}{}$

H. 64% = 0.64 $\frac{64}{100}$

J. 640%

(H is circled)

3 Which of the following numbers, when multiplied by 4^3, is equal to 1?

A. 4^{-3} $4 \times 4 \times 4 =$

B. -4^{-3} $12\sqrt{} = 48$

C. 4^3

D. -4^3 -2 $-4 \times -4 \times -4 = 64$

(A is circled)

4 What is another way to express 74,000?

F 27^2

G 34^2

H 7.4×10^3

(J) 7.4×10^4

Student work: $27 + 27 = 729$; $34 \times 34 = 1156$; $7.4,000$

5 Which of the numbers below is the greatest?

A $\sqrt{121} = 11$

B $\dfrac{121}{12} = 10$

(C) $1.21 \times 10^3 = 1210$

D 12% of 121 = $\dfrac{12}{121} = 1.9 =$ [2]

Student work: $\dfrac{12}{100}$, $\dfrac{121}{1}$

6 An architect found the height of a doorway on a blueprint to be $\sqrt{50}$ feet. Which point is closest to $\sqrt{50}$ on the number line?

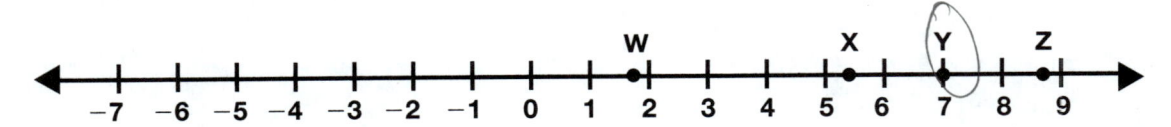

Number line with points W at 2, X at 5, Y at 7, Z at 8.

F W

G X

(H) Y

J Z

Student work: $\sqrt{50} \approx 7.0$

Chapter 1: Numbers and Operations 29

7 Which value is the greatest?

A 3^4

B 2^5

C 5^3

D 6^2

8 A building contractor is using a wire and a pulley to lift materials to the roof of a building.

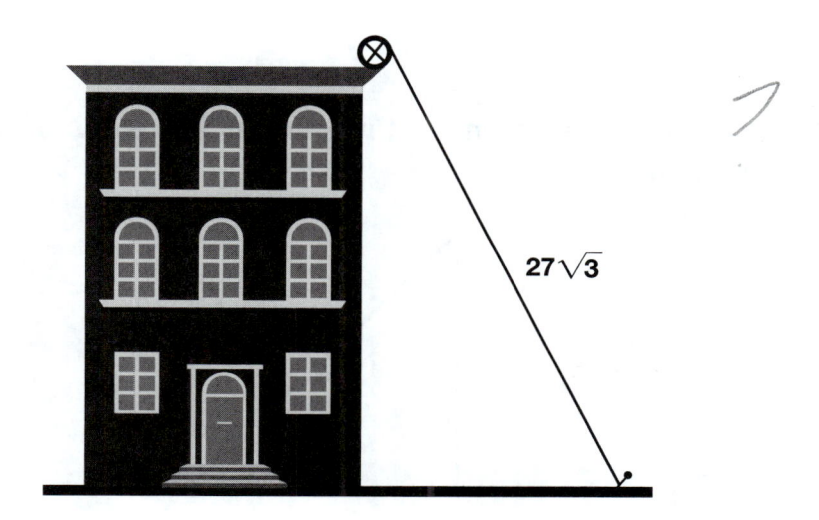

The contractor used the Pythagorean theorem to determine that the length of the wire is $27\sqrt{3}$. Which of the following numbers is closest to the length of the wire?

F 45 feet

G 50 feet

H 55 feet

J 60 feet

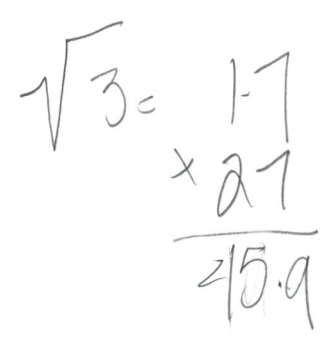

9 Which value is equal to 0.80?

A $\frac{1}{3} = 0.\overline{33}$

B $\frac{2}{3} = 0.\overline{66}$

C $\frac{4}{5} = 0.8$ ⟵ circled

D $\frac{6}{7} = 0.85$

10 The measurements, in centimeters, of the sides of a right triangle are shown below.

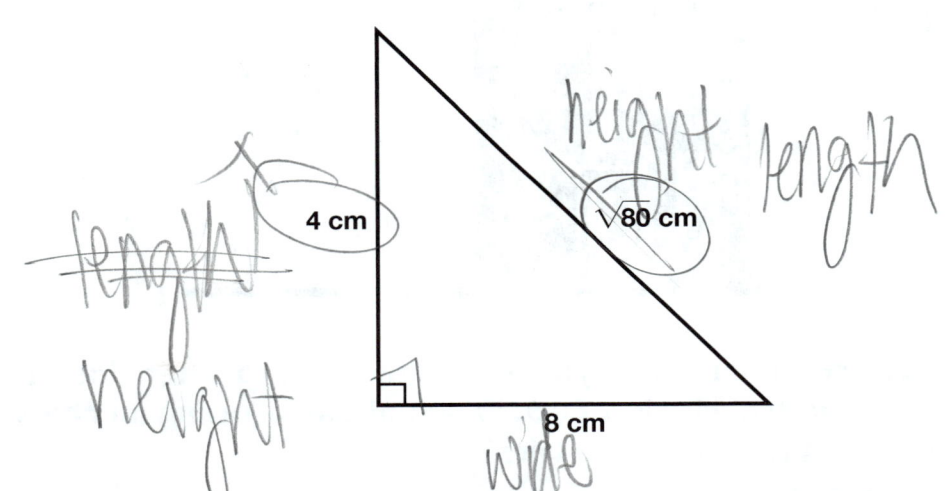

[Triangle with sides 4 cm, √80 cm, and 8 cm; right angle marked. Handwritten labels: "length", "height", "wide" around the triangle.]

Which of the following is closest to the length of the hypotenuse of this right triangle?

F 4

G 8

H 9 ⟵ circled

J 80

$\sqrt{80} = 8.9$

Chapter 1 Answer Explanations

Let's Review 1: Numbers and Operations

1. B
You need to move the decimal point in 21,000,000 seven places to get to 2.1.

2. F
If you press the square root key on your calculator and then the number 1, 4, 4, you'll see that the square root of 144 is 12. So, 12^2 is the correct answer. Or you can see that F is the correct answer by multiplying 12 by itself on your calculator to get 144.

3. B
You can answer this question two different ways. You can use your calculator to divide the denominator of each fraction into the numerator and compare the results to .45. You can also answer this problem by converting .45 into a fraction. To do this, move the decimal point two places to the right and put this number (45) over 100. Then reduce the resulting fraction (the common factor is 5) and compare to the answer choices.

4. 25%
To convert a decimal to a percentage, move the decimal point two places to the right (to represent hundredths, the meaning of %), and insert a percent sign.

5. C
You can see that the square root of 32 is approximately 5.6 by pressing the square root key on your calculator and then the number 32. Point *Y* is closest to this number.

Chapter 1 Practice Problems

1. C

To solve this problem, you can eliminate answer choices A and B, since they are obviously smaller than the numbers in answer choices C and D (the numbers as well as the exponents are smaller). Use your calculator to find the correct answer between choices C and D.

2. H

Remember that to convert a decimal into a percentage, move the decimal point two places to the right and write the percent sign.

3. A

The number 4^{-3} multiplied by 4^3 is 1. $4^{-3} \times 4^3 = 4^0 = 1$.

4. J

The correct answer choice to this question uses scientific notation. You need to move the decimal point from the end of 74,000 four places to the left to get between 7 and 4, so the exponent on 10 must be 4.

5. C

You might be able to answer this question by sight. If you move the decimal point in 1.21 three places to the right, the number is 1,210. All the other numbers are less than 121.

6. H

The square root of 50 is approximately 7.07. Point Y is closest to this number.

7. C

You can use your calculator to quickly figure out the answer to this question. The number 5^3 is 125, and this is the greatest number.

8. F

The square root of 3 is about 1.7. If you multiply this by 27, it is about 45.9.

9. C

If you convert the decimal 0.80 to a fraction, you get $\frac{80}{100}$. When you reduce this number, you get $\frac{4}{5}$.

10. H

The square root of 80 is about 8.9. Answer choice H is closest to this number.

Chapter 2
Equations and Inequalities

$2(x+y)=2x+2y$
$7y>30$

Objectives

Objective 1: The student will describe functional relationships in a variety of ways.

A(b)(1) **Foundations for functions**. The student understands that a function represents a dependence of one quantity on another and can be described in a variety of ways.

(A) The student describes independent and dependent quantities in functional relationships.

(C) The student describes functional relationships for given problem situations and writes equations or inequalities to answer questions arising from the situations.

(D) The student represents relationships among quantities using [concrete] models, tables, graphs, diagrams, verbal descriptions, equations, and inequalities.

Objective 2: The student will demonstrate an understanding of the properties and attributes of functions.

A(b)(3) **Foundations for functions**. The student understands how algebra can be used to express generalizations and recognizes and uses the power of symbols to represent situations.

(A) The student uses symbols to represent unknowns and variables.

(B) Given situations, the student looks for patterns and represents generalizations algebraically.

A(b)(4) **Foundations for functions**. The student understands the importance of the skills required to manipulate symbols in order to solve problems and uses the necessary algebraic skills required to simplify algebraic expressions and solve equations and inequalities in problem situations.

(B) The student uses the commutative, associative, and distributive properties to simplify algebraic expressions.

In this chapter, you'll learn how to answer questions about algebraic expressions and equations on the TAKS. Often you'll be asked to choose the expression or equation that could be used to solve a real-life problem. Some expressions and equations have more than one variable, and some are displayed in table form. You'll learn about expressions and equations in this chapter.

An inequality shows the relationship between two values. You'll also learn about inequalities and patterns in this chapter.

Order of Operations

Some questions on the TAKS will involve more than one operation. To answer these questions correctly, you have to perform operations in the correct order. Look at this problem:

$$2 + 8 \times 3$$

If you solve the problem from left to right, the answer is $10 \times 3 = 30$.

However, if you multiply first, $(8 \times 3) + 2$, the answer is $24 + 2 = 26$. This example uses the correct order of operations.

Chapter 2: Equations and Inequalities

Perform operations in this order:

1. Perform operations inside parentheses first. Within the parentheses, use the order in the steps below.
2. Next, perform powers from left to right.
3. Then multiply or divide from left to right.
4. Last, add or subtract from left to right.

Properties

A mathematical **property** can be thought of as a rule. Three commonly used basic mathematical properties are the commutative, associative, and distributive properties, as discussed next.

Commutative Property

The **commutative property of addition** says that quantities can be added in any order. Likewise, the **commutative property of multiplication** says that quantities can be multiplied in any order. In other words, the quantities can be moved around, or commuted, and the result will not change.

When the commutative property is applied to addition, for example, 3 + 6 is the same as 6 + 3; they both equal 9.

Sometimes letters, or **variables**, are used instead of numbers, the commutative property looks like this:

$$x + y = y + x$$

When the commutative property is applied to multiplication, for example, 2×6 is the same as 6×2; they both equal 12. Using variables, the commutative property looks like this:

$$x \times y = y \times x, \text{ or } xy = yx$$

The following problem involves the commutative property.

Daryl plans to spend $6.00 per person for a surprise party for his mother. Ten people are coming to the party. His total cost can be expressed as $6.00 × 10.

Use the commutative property to write an equivalent expression.

A $6.00 + 10
B 10 × $6.00
C $6.00 × 10
D $6.00 ÷ 10

To correctly answer this question, you need to know that the commutative property of multiplication means that two quantities can be multiplied in any order. The only answer that shows a different order for multiplication of the two numbers is choice B.

Note that the commutative property does not hold for subtraction (for example, $8 - 2$ is not the same as $2 - 8$) or division (for example, $\frac{6}{3}$ is not the same as $\frac{3}{6}$).

Associative Property

The **associative property** of addition says that three or more quantities being added can be combined, or associated, with each other in any order and the result will not change. Similarly, the associative property of multiplication says that three or more quantities being multiplied can be combined, or associated, with each other in any order and the result will not change.

Chapter 2: Equations and Inequalities

When the associative property is applied to addition, for example, 5 + 2 + 3 can be written as (5 + 2) + 3 or 5 + (2 + 3), where the parentheses indicate which addition to do first. Both ways add to 10.

Using variables, the associative property for addition can be written:

$$a + b + c = (a + b) + c = a + (b + c).$$

When the associative property is applied to multiplication, for example, 2 × 3 × 4 can be written as (2 × 3) × 4 or 2 × (3 × 4), where the parentheses indicate which multiplication to do first. Both ways give the same answer, 24.

Using variables, the associative property for multiplication can be written as:

$$(a \times b) \times c = a \times (b \times c), \text{ or } (ab)c = a(bc)$$

The following problem involves the associative property.

During her first year on the basketball team, Sarah scored x points. During her second year, she scored 65 points, and during her third year, she scored 75 points. Her total points for the three years could be expressed as $x + (65 + 75)$.

Use the associative property to write an equivalent expression.

A. $x = 65 + 75$

B. $65x + 75x$

C. $x(65 + 75)$

D. $(x + 65) + 75$

This question asks you about the associative property of addition. Multiplication should not be involved, so you can eliminate answer choices B and C. Answer choice A does not add the three values. Answer choice D is correct.

Note that the associative property does not hold for subtraction or division.

Distributive Property

The **distributive property** is used when an expression involving addition or subtraction is multiplied by a quantity, as in 5(2 + 3). According to the distributive property the quantity 5 multiplies each of the terms in the parentheses, or is "distributed" to each term in the parentheses. Thus, $5(2 + 3) = (5 \times 2) + (5 \times 3) = 10 + 15 = 25$. You may notice that if you simply used the order of operations, which says to do the operation in parentheses first, you would get 5(5) = 25, which seems easier. However, if variables are involved in an expression, the distributive property can be very useful.

For example, $5(x + y - 2) = 5x + 5y - 10$ uses the distributive property.

Expressions

In algebra, letters often stand for numbers that need to be determined. You just learned that these letters are called **variables**. Any number in front of a variable means that the variable will be multiplied by that number, called a **coefficient**. For example, $2x$ means that x will be multiplied by 2. Any single variable, such as x or y, has a 1 before it even though the coefficient 1 is not written. In other words, $x = 1x$ and $y = 1y$.

An algebraic **expression** has at least one variable, but not an equal sign. Parentheses are often used in algebraic expressions. So far in this chapter, you learned about the order of operations and three basic arithmetic properties which can be used to simplify expressions. For example, to simplify the expression

$$4(3x) + x$$

begin by using the order of operations, and multiplying out the first term:

$$12x + x$$

Now simplify it even further by adding. Remember that x is the same as $1x$.

$$13x$$

Chapter 2: Equations and Inequalities   41

Questions on the TAKS may ask you to choose the correct expression based on a given situation. Read this question:

Mabel bought 12 pencils for 20 cents each, 6 pens for 50 cents each, 2 erasers for 50 cents each, and 5 sheets of construction paper for 20 cents each. Which expression would enable Mabel to find out how much money she spent?

A $12 \times 6 \times 2 \times 5$

B $17(20) + 8(50)$

C $(17 + 20) \times (8 + 50)$

D $(12 + 20) \times (6 + 50) \times (2 + 50) \times (5 \times 20)$

The correct answer is B.

Add the costs, item by item:

$$12(20) + 6(50) + 2(50) + 5(20)$$

Then combine like items (items with the same cost) to get:

$$17(20) + 8(50)$$

Let's try one more:

Abraham uses the expression $7x + 10.5y$ to determine the amount he earns at a pay rate of seven dollars an hour plus time and a half for overtime. One week he worked 40 regular hours, plus 2 hours of overtime. Write an expression to determine how much Abraham earned.

The information in this problem gives you the values to substitute for x and y. Substitute 40 for x, the number of regular hours Abraham worked and substitute 2 for y, the number of overtime hours Abraham worked. This expression would help Abraham determine how much money he would earn in a week.

$$7(40) + 10.5(2)$$

Equations

An algebraic **equation** is a statement that says two values, or expressions, are equal. You can spot an equation easily because it has an equal sign, which separates the two sides of the equation. Because the two sides of an equation are equal, whatever you do to one side, you must do to the other side. You will often use inverses, discussed in Chapter 1, when working with equations.

For example, look at this equation:

$$t + 45 = 100$$

In this equation, t is the variable, which represents the unknown quantity. You can solve the equation, and find the value of t, by subtracting 45 from each side of the equation. Remember that the additive inverse of 45 is -45.

$$t + 45 - 45 = 100 - 45$$
$$t = 55$$

Some questions on the TAKS will ask you to choose the correct equation based on a situation. Read this problem:

Terry's take-home (net) pay is his gross pay minus the $175 his employer deducts each week for taxes. His net pay is $500 a week. Write an equation that could be used to find Terry's gross pay.

Terry's gross pay is the unknown variable, or x. This equation could be used to determine his gross pay:

$$x - \$175 = \$500$$

To solve this equation and determine Terry's gross pay, isolate or get x, the variable, on one side by itself by adding $175 to each side:

$$x - \$175 + \$175 = \$500 + \$175$$
$$x = \$675$$

Chapter 2: Equations and Inequalities 43

Some questions on the TAKS will involve equations with more than one variable. For an equation with multiple variables, such as $2x + y = 10$, you solve for one variable in terms of another. To solve this equation for y in terms of x, move everthing except x to the other side of the equation.

$$2x + y = 10$$

Subtract $2x$ from each side:

$$2x + y - 2x = 10 - 2x, \text{ or}$$
$$y = 10 - 2x$$

Let's try another one. This time solve for b in terms of a.

$$4b = 16a$$

Divide each side by 4. Remember that division is the inverse of multiplication.

$$\frac{4b}{4} = \frac{16a}{4}$$
$$b = 4a$$

Equations with more than one variable are sometimes written in table form. The table below shows the values of x and y for the equation $3x - y = 0$.

x	y
2	6
3	9
4	12
5	15
6	?

What value makes the equation true when $x = 6$?

If you substitute 6 for x, the equation looks like this: $3(6) - y = 0$. You could probably figure this out in your head. However, the equation can also be solved this way:

$$18 = 0 + y, \text{ or } 18 = y.$$

Inequalities

An **inequality** links two expressions by using these signs:

>	greater than
<	less than
≥	greater than or equal to
≤	less than or equal to
≠	not equal to

Questions about inequalities might also use words, such as *greater than, less than, between, at least,* or *at most.*

Solving an inequality is very similar to solving an equation. Look at this inequality:

$$3y > 21$$

Get *y* on a side by itself by dividing both sides by 3:

$$\frac{3y}{3} > \frac{21}{3}$$

$$y > 7$$

This means *y* can be any number that is greater than 7.

Even though working with inequalities is similar to working with equations, there is one important difference. If you multiply or divide both sides by a negative number, the inequality sign reverses. For example:

Solve $-3y \geq 21$ by dividing both sides by -3:

$$\frac{-3y}{-3} \geq \frac{21}{-3}$$

$$y \leq -7$$

Notice the inequality symbol turned around.

Patterns

Patterns appear throughout the field of mathematics. To answer questions about patterns, you will have to choose the next number, letter, or picture in a sequence. Look at this simple pattern:

A, B, C, D, E, A, B, C, D, E, A, B…

Which letter comes next? Did you guess C? Now look at the pattern again. There are 5 letters in the pattern. Which letter would be the 20th? To determine this, you would divide by 5. The number 5 goes into 20 evenly, 4 times. Therefore the 20th letter in the pattern would be the same as the 5th, the letter E.

Let's Review 2: Equations

Complete each of the following questions. Use the Tip below each question to help you choose the correct answer. When you finish, check your answers with those at the end of Chapter 2.

1 A sporting goods store had 52 sweatshirts at the beginning of a sale. If *y* represents the number the sweatshirts sold during the sale, which expression shows the number of sweatshirts remaining?

A $y - 52$

B $52(y)$

C $52 - y$

D $y + 52$

TIP: Remember that the store had 52 sweatshirts before the sale, so the amount sold would have to be subtracted from this amount.

2 The number of plain white straws Cara has is shown by the expression $3x + 4$, with *x* representing the number of striped straws. If Cara has 10 striped straws, how many plain white straws does she have?

F 16

G 30

H 34

J 70

TIP: Evaluate the expression like this: $3(10) + 4$.

Chapter 2: Equations and Inequalities   47

3 Morgan's age is shown by the expression a + 3, where a represents Andrea's age. If Andrea is 9, how old is Morgan?

A 6
B 9
C 12 *(circled)*
D 15

Handwritten: 9+3, 9+3=12

TIP Add 9 and 3.

4 Solve for y, if $\frac{3}{y} = \frac{1}{2}$.

F 2
G 3
H 4
J 6 *(circled)*

Handwritten: $\frac{3}{4} = \frac{1}{2}$, 6, $\frac{3 \div 3}{6 \div 3} = \frac{1}{2}$

TIP What fraction with 3 as a numerator can be reduced to $\frac{1}{2}$?

5 A party planner charges a flat fee of $100 to plan a birthday party and an additional $15 per guest. If n = the number of guests and c = total charges, which of the following shows how to determine the total charges?

A c = $100 + $15n *(circled)*
B c = $100n + 15
C $c = n + \frac{100}{15}$
D $c = \frac{100n}{15}$

Handwritten: $100 + $15, c = $100 + $15N

TIP The variable n is the number of guests.

48 TAKS   Mathematics

6 The table below shows values for *x* and *y* for the equation $x^2 - y = 1$.

x	y
2	3
3	8
4	15
5	24
6	35
7	48
8	?

What value of *y* makes this equation true when *x* = 8?

Be sure to use the correct place value when marking the grid.

TIP Substitute 8 into the equation. Determine what you need to subtract to get 1.

7 A salesperson's total salary includes a base pay of $800 a month plus 2.5% of the monthly sales. If *x* = sales per month and *y* = total salary, which of the following shows how to determine the total salary for any month?

A $800 = y + .025x$

B $y = \$800 + .025x$

C $y = \$800 \times .025x$

D $y = \$800x \times .025$

TIP Remember that 2.5% of the sales will be added to $800.

Chapter 2: Equations and Inequalities   49

8 Given the inequality $3y < 18$, solve for y.

F $y = 6$

G $y < 6$

H $y > 6$

J $y \leq 6$

TIP: The sign should be the same as the sign in the inequality.

9 Alice made this design on a wall in her room.

If the bottom flower is the first row, how many flowers will be in the seventh row?

A 7

B 14

C 64

D 128

TIP: Look carefully at the relationships between each two rows.

50 TAKS   Mathematics

Chapter 2 Practice Problems

Complete each of the following practice problems. Check your answers at the end of this chapter. Be sure to read the answer explanations!

1. A real estate agent is paid each time he sells a house. He is paid a flat fee of $1,000 and a commission of 4% of the sale price of the house. If s = sale price of the house and c = the agent's total payment, which of the following equations shows how to determine the real estate agent's payment?

- A $c = \$1,000 + .004s$
- B $c = \$1,000 - .004s$
- C $c = \$1,000 + .04s$
- D $c = \$1,000 - .04s$

2. What is the value of x if $3x + 3 = 12$?

- F 2
- G 3
- H 4
- J 9

3. The number of cats at an animal shelter is shown by the expression $2y - 5$, with y representing the number of dogs. If the shelter has 125 dogs, how many cats does it have?

- A 130
- B 245
- C 250
- D 255

Chapter 2: Equations and Inequalities   51

4 The table below shows values of *x* and *y* for the equation $x^3 = y$.

x	y
2	8
3	27
4	?
5	125
6	216
7	343

[handwritten: 4×4×4 ↓ 16×4=64]

What value of *y* makes this equation true when *x* = 4?

Be sure to use the correct place value when using the answer grid.

5 At a flea market, a vendor sold 1 handmade quilt and 2 antique plates for less than $100. If *q* represents the selling price of the quilt and *p* represents the selling price of one plate, which inequality could be used to show the possible amounts of money the vendor made?

A $q + p < \$100$

B $q + 2p < \$100$ *(circled)*

C $q + p \leq \$100$

D $q + 2p \leq \$100$

[handwritten: q + ; q + 2p < $100]

6 Look at the letter pattern below.

ABCDABCDABCDABCD

Which letter would be in the twentieth place?

F A

G B

H C

J D

7 The drill team at Karen's school is planning a half-time show. The figure below represents the pattern the team wants to create.
The pattern has an innermost ring of 12 members of the drill team. Each additional ring needs 4 more members than the previous ring.

If the half-time show consists of five rings formed according to this pattern, what will be the total number of drill team members needed to form all five rings?

A 24

B 29

C 48

D 100

Chapter 2: Equations and Inequalities   53

Chapter 2 Answer Explanations

Let's Review 2: Equations

1. C

If the store had 52 sweatshirts before the sale and y represents the unknown number of sweatshirts sold during the sale, an expression that could be used to find the number of sweatshirts remaining is $52 - y$.

2. H

To simplify the expression $3x + 4$, put 10 in place of x: $3(10) + 4 = 34$. Cara has 34 plain white straws.

3. C

If Andrea is 9 and Morgan is Andrea's age plus 3, Morgan is 12.

4. J

You may be able to solve this equation mentally. If not, multiply both sides of the equation by y to get $\frac{3}{y} \times y = \frac{y}{2}$, or $3 = \frac{y}{2}$. Then multiply both sides by 2 to get $2 \times 3 = 2 \times \frac{y}{2}$, or $y = 6$.

5. A

In this case, the number of guests, n, is the variable, along with the cost, c. The cost is $15 per guest times n plus the $100 flat fee. Answer choice A is correct.

6. 63

If you substitute 8 for x in the equation, $x^2 - y = 1$, you get $64 - y = 1$. You can probably see that $y = 63$ is the correct answer. If not, add y to both sides of the equation to get $64 - y + y = 1 + y$, or $64 = 1 + y$. Then subtract 1 from both sides to get $64 - 1 = 1 + y - 1$, or $63 = y$.

7. B

There are two variables in this equation, the salesperson's total salary and the monthly sales. The monthly sales, x should be multiplied by 2.5%, or .025, then added to the salesperson's base pay of $800.

8. G

To solve this inequality, get y on one side by itself:

$$3y < 18$$
$$\frac{3y}{3} < \frac{18}{3}$$
$$y < 6.$$

9. C

The number of flowers doubles in each row. So the number of flowers in the seventh row would be 64.

Chapter 2 Practice Problems

1. C

The percentage, 4%, equals .04, and this is multiplied by the sale price of the house, s, to determine the commission. The commission is added to the flat fee of $1,000 for the total payment.

2. G

To solve the equation $3x + 3 = 12$, isolate x on one side of the equation:

$3x + 3 - 3 = 12 - 3$, or $3x = 9$, and thus $x = 3$.

3. B

To evaluate this expression, substitute the number of dogs for y in the expression $2y - 5$. Then $2(125) - 5 = 245$.

Chapter 2: Equations and Inequalities 55

4. 64

Substitute 4 for x in $x^3 = y$. The number 4 cubed is 64.

5. B

There are two plates, so you need to put the number 2 in front of p, the variable representing plates. The total price must be less than $100.

6. J

There are four letters in this series and 4 goes into 20 evenly, so the pattern completes itself and the last letter is D.

7. D

Two steps are needed to solve this problem. First, determine how many drill team members will be in each ring. Then add the number of members in each ring. The numbers to be added are 12, 16, 20, 24, and 28.

Chapter 3

Functions, Part 1

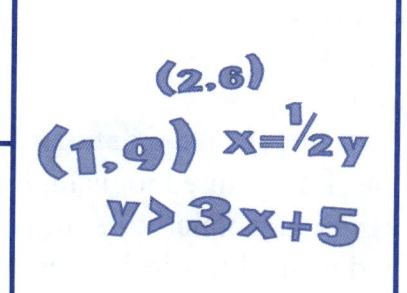

(2,6)
(1,9) $x = \frac{1}{2}y$
$y > 3x+5$

Objectives

Objective 1: The student will describe functional relationships in a variety of ways.

A(b)(1) Foundations for functions. The student understands that a function represents a dependence of one quantity on another and can be described in a variety of ways.

(B) The student [gathers and records data, or] uses data sets, to determine functional (systematic) relationships between quantities.

(E) The student interprets and makes inferences from functional relationships.

Objective 2: The student will demonstrate an understanding of the properties and attributes of functions.

A(b)(2) Foundations for functions. The student uses the properties and attributes of functions.

(C) The student interprets situations in terms of given graphs [or creates situations that fit given graphs].

A(b)(4) Foundations for functions. The student understands the importance of the skill required to manipulate symbols in order to solve problems and uses the necessary algebraic skills required to simplify algebraic expressions and solve equations and inequalities in problem situations.

(A) The student finds specific function values, simplifies polynomial expressions, transforms and solves equations, and factors as necessary in problem situations.

Objective 4: The student will formulate and use linear equations and inequalities.

A(C)(4) Linear functions. The student formulates systems of linear equations from problem situations, uses a variety of methods to solve them, and analyzes the solutions in terms of the situation.

(A) The student analyzes situations and formulates systems of linear equations to solve problems.

(B) The student solves systems of linear equations using [concrete] models, graphs, tables, and algebraic methods.

(C) For given contexts, the student interprets and determines the reasonableness of solutions to systems of linear equations.

In this chapter, you'll learn about relationships in which variables are related to one another. You'll learn about independent and dependent quantities, and you'll begin to learn about functions. You'll also expand your knowledge of inequalities, expressions, and equations in this chapter.

Dependent and Independent Variables

When the outcome of one event has no effect on another event, the events are not related to each other. Some events are related, however—one event directly affects the other. In this case, there is often a dependent and independent variable. The **independent variable** causes the change in the **dependent variable**. For example, imagine the results of a study regarding the amount of fruit people eat and the number of colds they get. The study found there is a relationship between the two: the more fruit a person consumes, the fewer colds he or she gets. In this case, the amount of fruit a person eats is the independent variable, or the cause. The number of colds a person gets is the dependent variable, or the effect.

You might be asked a question like this about independent and dependent variables on the TAKS:

A hot-chocolate vendor made a table showing the relationship between the daily low temperature and the number of cups of hot chocolate sold per day. What is the dependent quantity in this relationship?

A The daily low temperature

B The number of cups of hot chocolate sold per day

C All of the data in the table

D Cannot be determined

In this case, the daily low temperature is the independent variable. It causes the change in the dependent variable, the number of cups of hot chocolate sold per day. Answer choice B is correct.

Functional Relationships

In very simple terms, a **function** is a relationship between two variables. A function takes some sort of *input*, usually a number, and changes it in some way to produce a single *output*. All numbers in a function are changed in the same way.

Look at the numbers in the table below.

x	y
1	5
2	6
3	7
4	8

The numbers in this table represent a function. They are all changed in the same way. The numbers in this function can be represented by the equation, $y = x + 4$.

Sometimes the dependent variable, in this case y, is written in shorthand as $f(x)$, which means "function of x." So this same function can be written as $f(x) = x + 4$. In a function, x corresponds to only one value of y. The x and y values are called **coordinates**. In the above table, the coordinates are (1, 5), (2, 6), (3, 7), and (4, 8). To graph this function, you would plot these coordinates on a grid and then draw a line connecting them.

Not all coordinates represent a function, however. Consider these coordinates: (2, 3), (2, −2). These coordinates can't be part of a function because the 2, the x-coordinate, cannot produce both 3 and −2. In a function, x corresponds to only one value of y. When graphed, a function will pass the vertical line test. This means that the graph of a function will cross any line parallel to the y-axis only once. The graph below is a function:

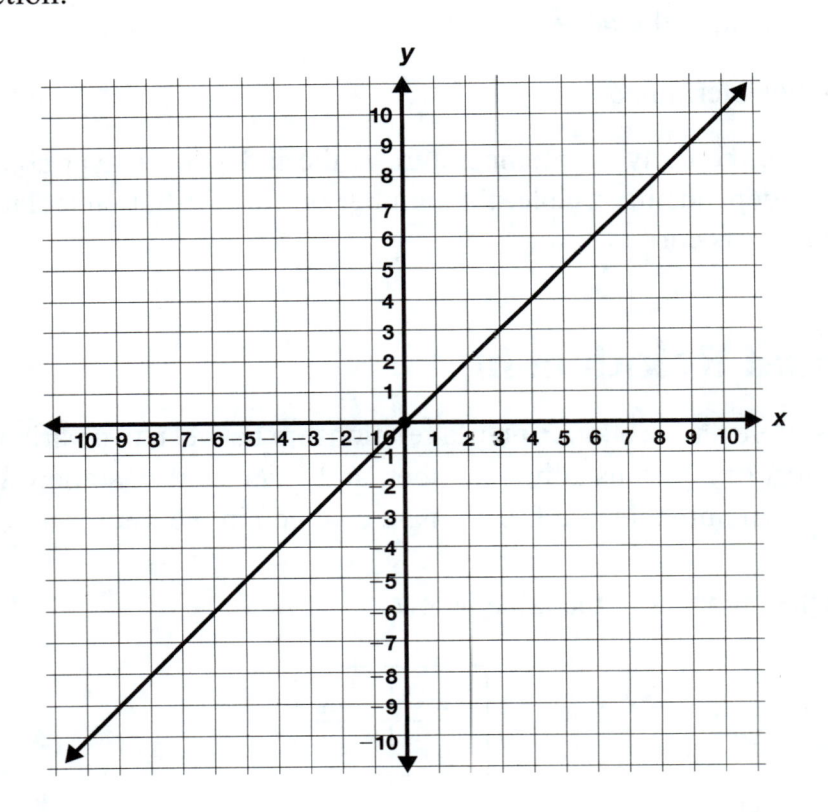

Chapter 3: Functions, Part 1

The following graph doesn't represent a function because it crosses the *y*-axis (and other vertical lines) more than once.

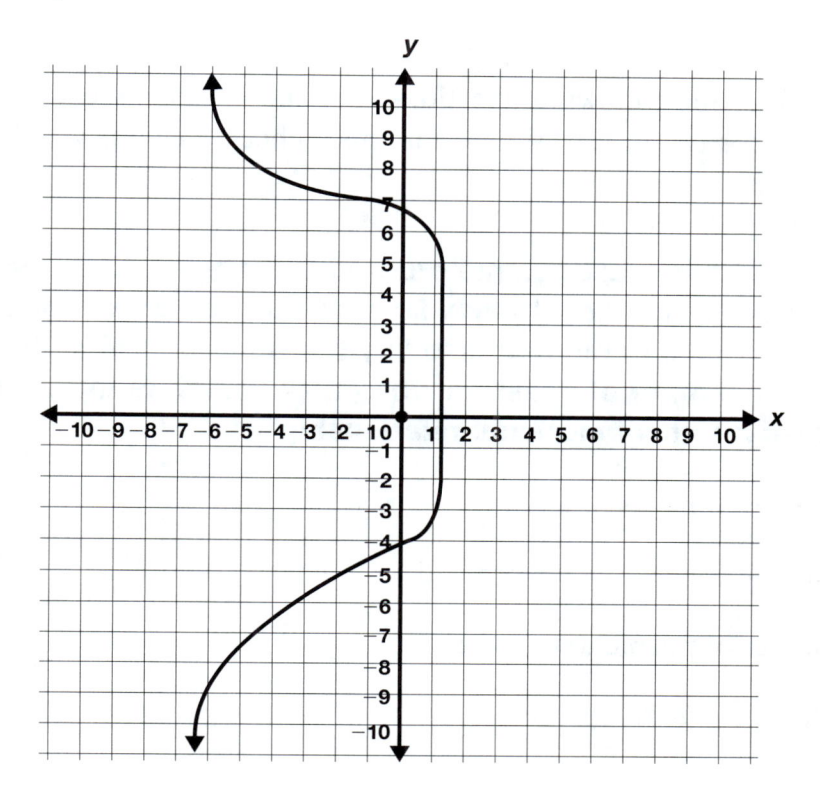

Let's Review 3: Functions

Complete each of the following questions. Use the Tip below each question to help you choose the correct answer. When you finish, check your answers with those at the end of Chapter 3.

1 For a birthday party, Jamie ordered 2 party favors for each guest and a box of 20 additional party favors to use as prizes for games. This relationship can be expressed by the function $f(s) = 2s + 20$, where s is the number of guests at the party. Which is the dependent quantity in this functional relationship?

A The number of boxes ordered

B The number of party favors ordered

C The number of people at the party

D The number of games played

TIP

Remember that the independent quantity or variable causes the change in the dependent quantity or variable. This question is asking for the dependent quantity.

2 Which graph is not a function?

F

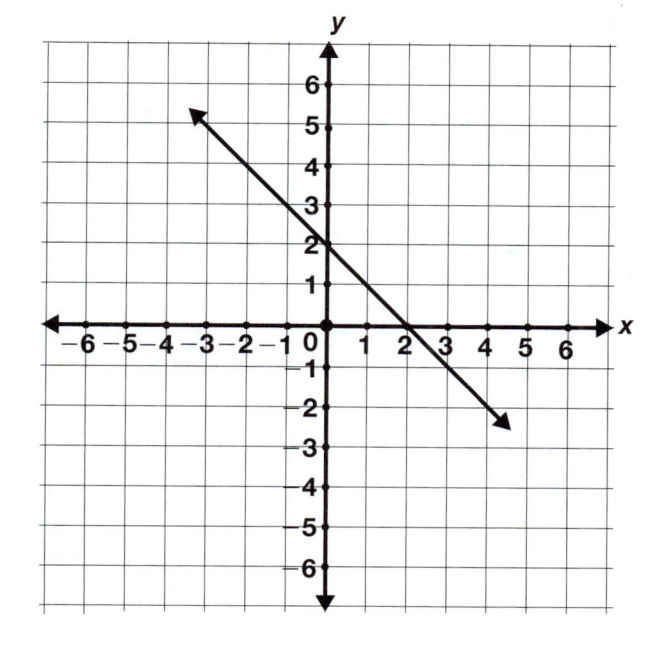

G

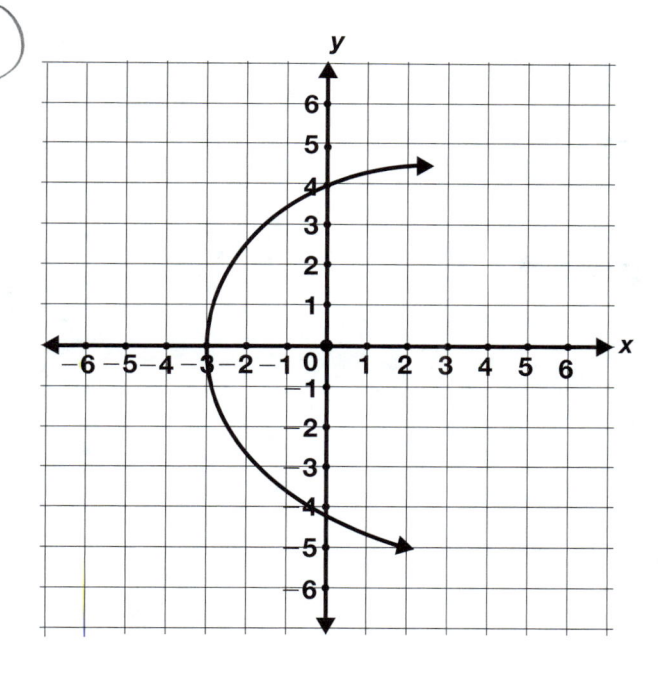

64 TAKS *Mathematics*

H

J

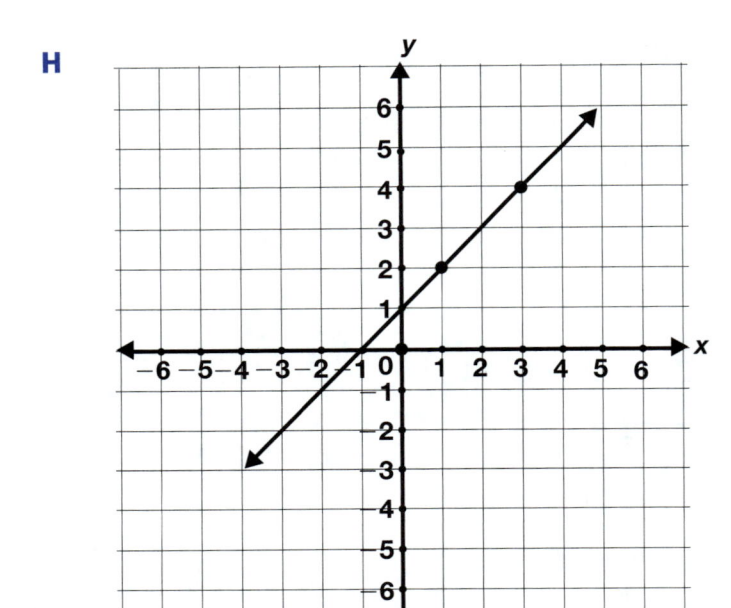

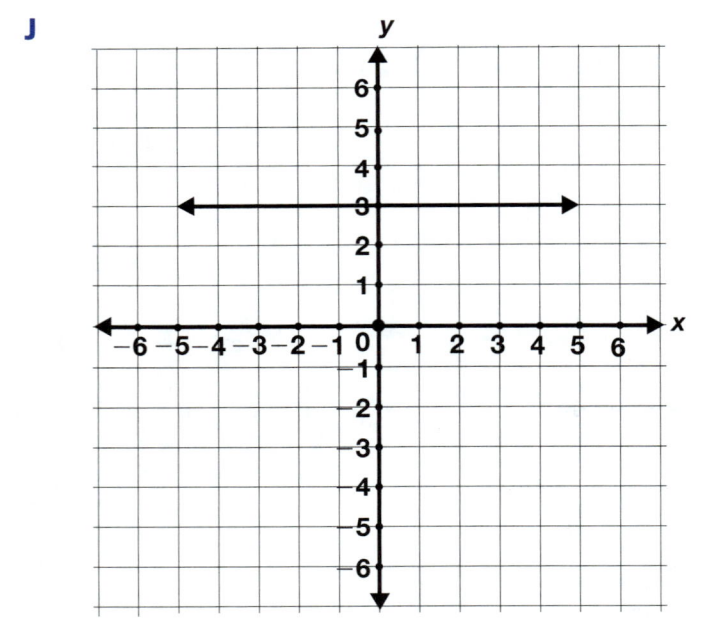

Remember the vertical line test.

3 The table below gives values for the function 2x = y. Fill in the missing values.

x	y
2 ×2	4
3 ×2	6
4 ×2	8
5 ×2	? 10
6 ×2	? 12

TIP

Be sure to multiply x by 2 to get y.

Graphing Inequalities

In the previous chapter of this book, you learned that an inequality links values that may or may not be equal. For example, $y > 3x - 1$ is an inequality. You can find the values for the variables in inequalities just as you would for an equation with an equal sign. The basic line for an inequality graph is found the same as the line for a function, as if the inequality has an equal sign. Look at these values for the boundary of the inequality $y > 3x - 1$:

x	y
1	3
2	6
3	9
4	12

The boundary line is drawn solid if these signs are used: $\leq$ and $\geq$, and the line is drawn dashed if these signs are used: $<$ and $>$.

After drawing the line, choose a point on one side of the line and substitute the coordinates of this point into the inequality. If they make the inequality true, all the points on that side of the line will make the inequality true, so shade the grid on that side of the line. If the coordinates make the inequality false, shade the grid on the other side of the line.

This is a graph of the inequality $y > 3x - 1$:

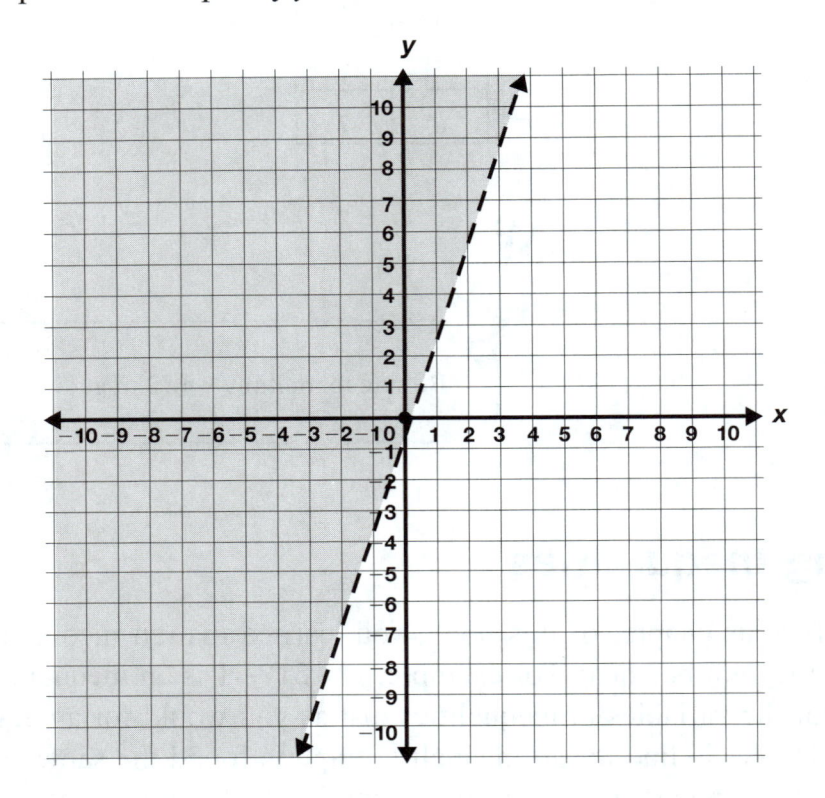

First, graph the line $y = 3x - 1$ as a dashed line because it is not included in the $>$ inequality. Then choose a point on either side of this line and check its truth in the original inequality. Let's say you choose the coordinate $(0, 3)$, which is above the line. Substitute these values into the inequality to get $3 > 3(0) - 1$, or $3 > -1$. Since this is true, you shade the area that contains $(0, 3)$, which is all of the grid above the line.

Selecting Expressions and Equations

Some TAKS questions will ask you to choose an equation or expression to represent a relationship with an independent and a dependent variable. For example, consider this scenario:

Chapter 3: Functions, Part 1

Ronnie decided to invest the money he earned during the last year. He invested $7000 of the money at an annual rate of 3% and the rest of the money, x, at an annual rate of 2.5%. Which equation describes y, the total amount of interest earned from both investments during the first year?

This scenario can be represented by this equation: $y = 0.03(7000) + 0.025x$. Note that the percentages have been converted to decimals and that x is the independent variable, since we don't know how much money Ronnie invested at a rate of 2.5%.

Linear Equations

A **linear equation** is an equation that forms a straight line when graphed. When you graph an equation, you substitute values for variables which are unknown quantities. On the TAKS, linear equations usually have more than one variable. Look at the equation shown here:

$$x + y = 4$$

If you listed the numbers that could be substituted for the variables x and y, your list might look like this:

x	y
0	4
1	3
2	2
3	1
4	0

If you graphed these numbers and drew a line, it would look like this:

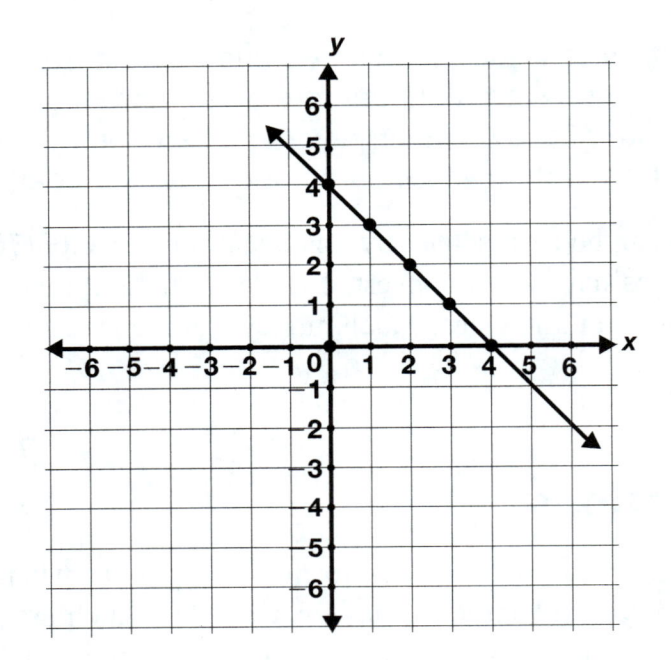

Some test questions will ask you to choose the correct graph of an equation. For others, you will be given a graph and asked to choose the correct equation.

Chapter 3: Functions, Part 1

Let's Review 4: Linear Equations

Complete each of the following questions. Use the Tip below each question to help you choose the correct answer. When you finish, check your answers with those at the end of Chapter 3.

1 Melanie is 2 years older than her sister Rachel. Which expression represents Rachel's age, if *x* represents Melanie's age and *y* represents Rachel's age?

A $y = 2x$

B $y = x - 2$

C $y = \dfrac{1}{2}x$

D $y = \left(\dfrac{2}{1}\right)x$

TIP: Remember that Melanie is older than Rachel.

2 Which of the following is a graph of the inequality $y > x - 1$?

F

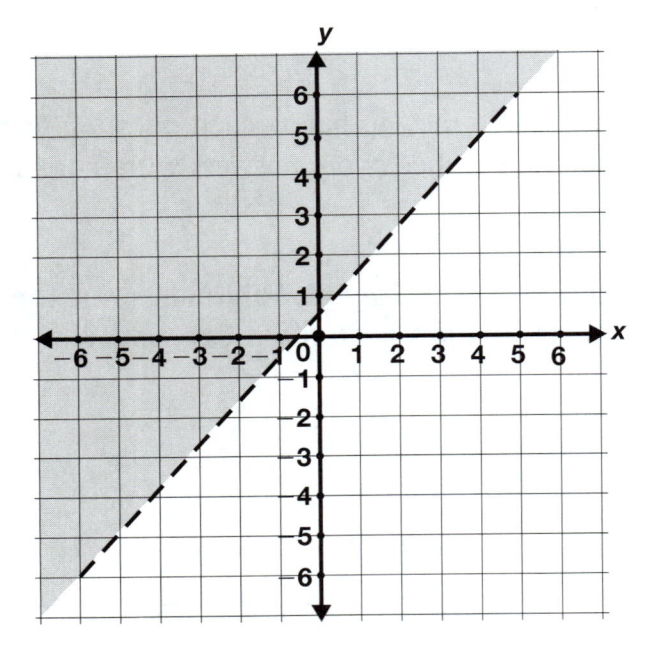

G

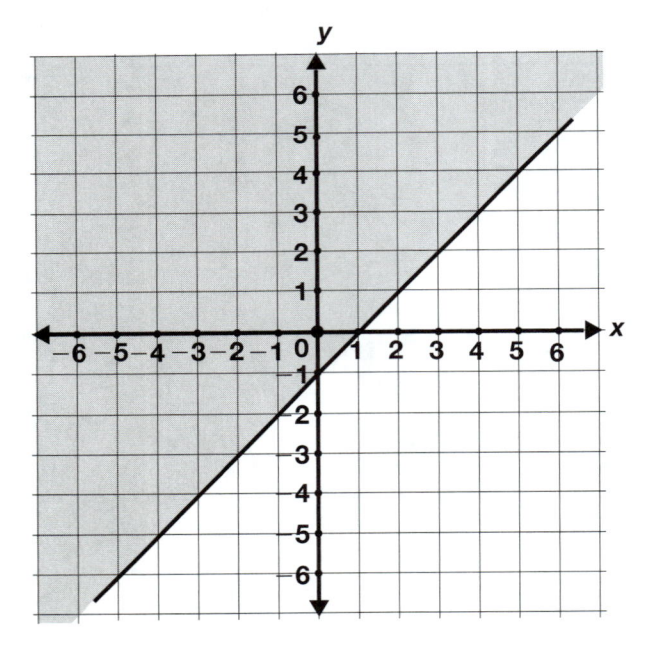

Chapter 3: Functions, Part 1   71

H

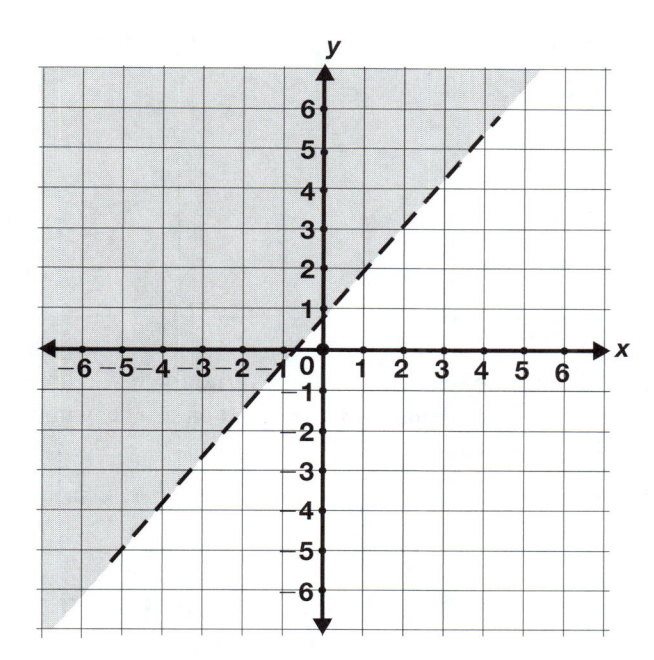

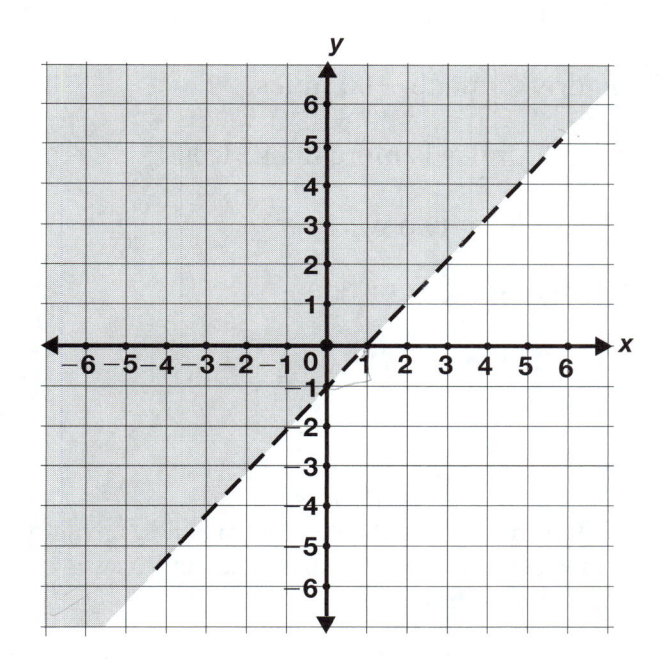

J

$y > x - 1$

> **TIP**
> Substitute values into the equation $y = x - 1$ to find the line. Then choose the correct graph for $y > x - 1$.

3 If *y* varies directly with *x*, and *y* is 12 when *x* is 6, which of the following can represent this situation?

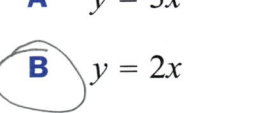

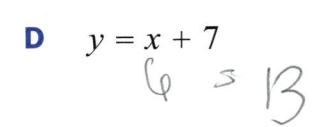

A $y = 3x$

B $y = 2x$

C $y = \left(\frac{1}{2}\right)x$

D $y = x + 7$

TIP: Choose the relationship that is true for $x = 6$ and $y = 12$. The equation will appear as $y = kx$, where k is a number.

4 The total cost, *c*, of leasing a car for a week can be expressed by the equation $c = 250 + 2m$, where *m* is the number of miles the car is driven. Which statement is true based on the information given?

F It costs more to lease the car if you drive it under 100 miles.

G The total cost of leasing this car for 500 miles is more than $1,000.

H It costs less to lease the car if you lease it for two weeks.

J The total cost of leasing this car for 500 miles is $1,000.

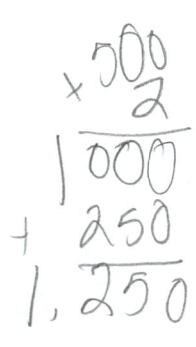

TIP: Eliminate answer choices that you know are untrue, then determine whether the remaining answer choices are true.

Chapter 3 Practice Problems

Complete each of the following practice problems. Check your answers at the end of this chapter. Be sure to read the answer explanations!

1 A bus traveled at a maximum speed for several hours. When it began to rain, the bus slowed down for a half hour and then returned to its maximum speed. Which of the following graphs best represents this information?

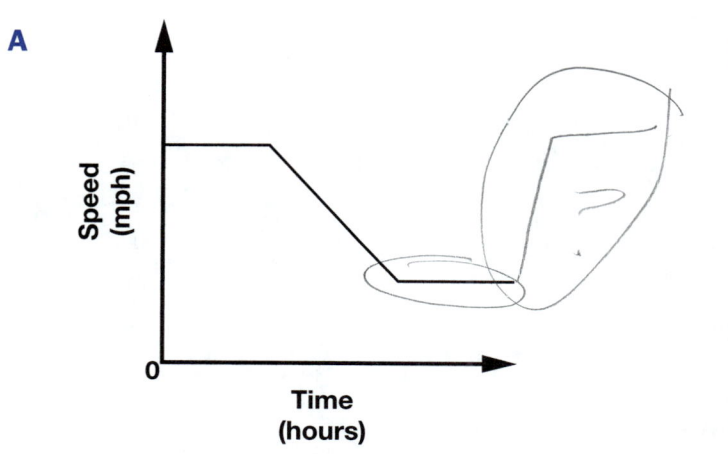

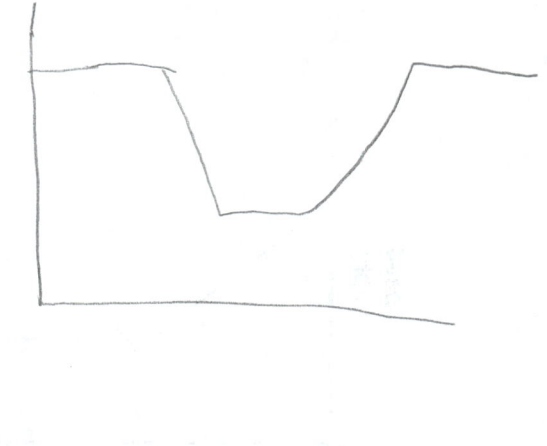

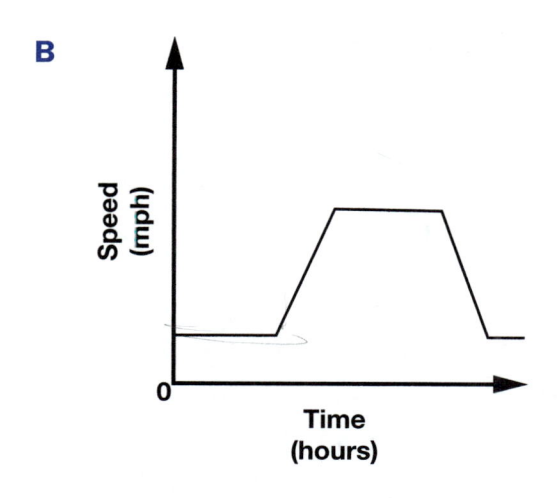

A

B

C

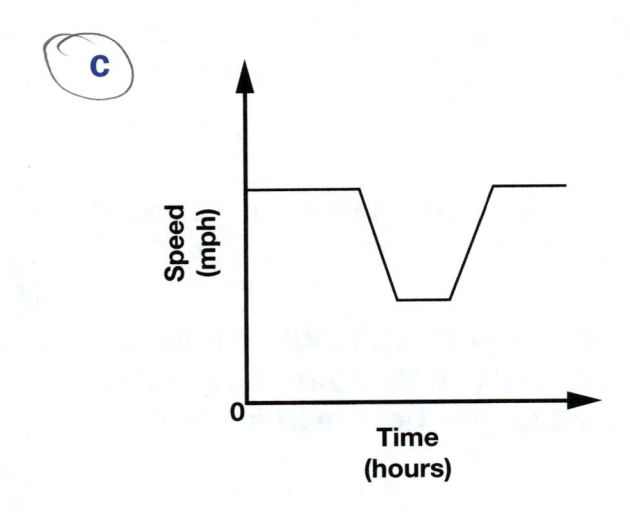

Speed (mph) vs Time (hours)

D

Speed (mph) vs Time (hours)

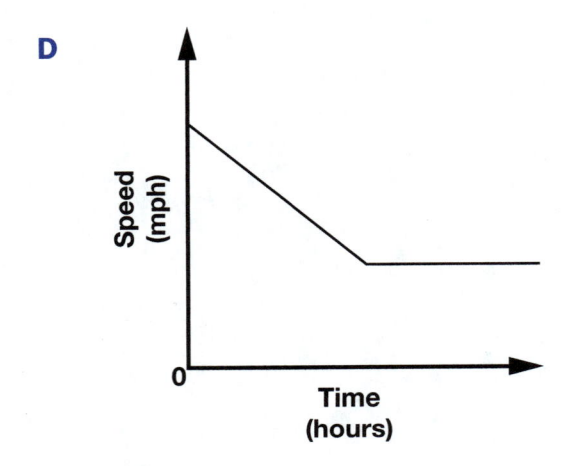

2 Which graph represents the equation $y \leq x + 2$?

F

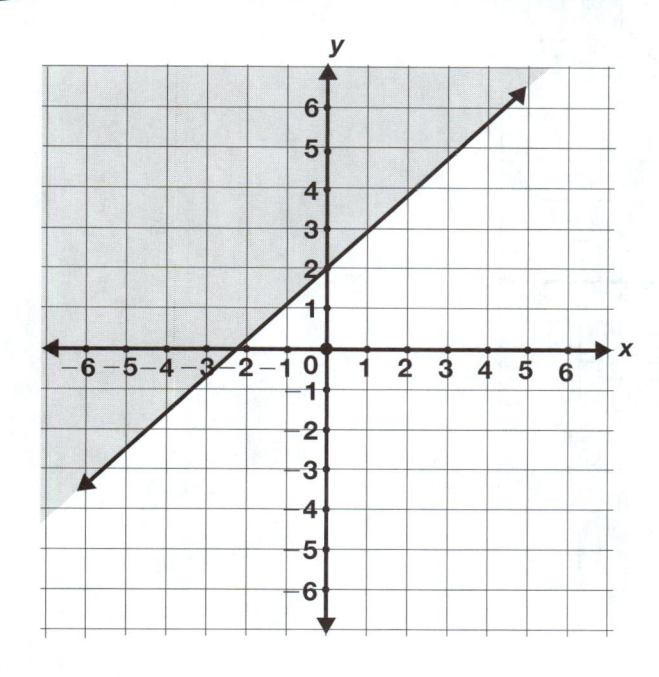

$y < x+2$

(G)

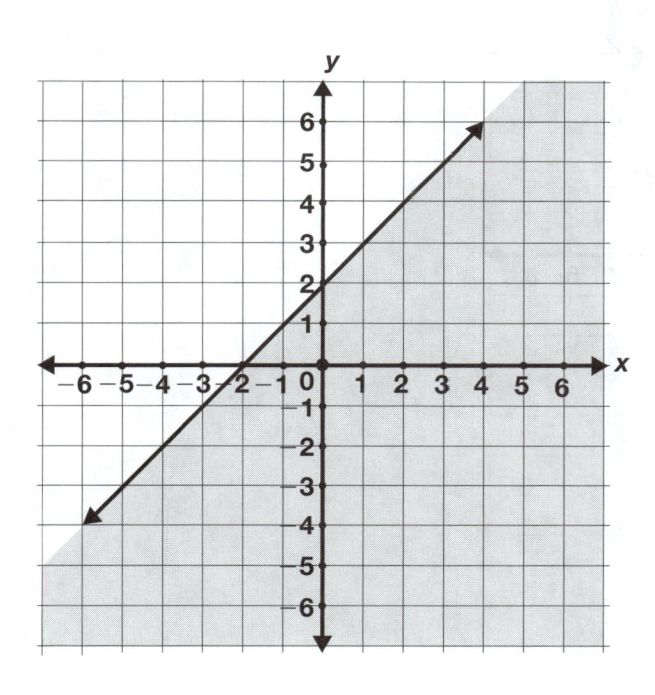

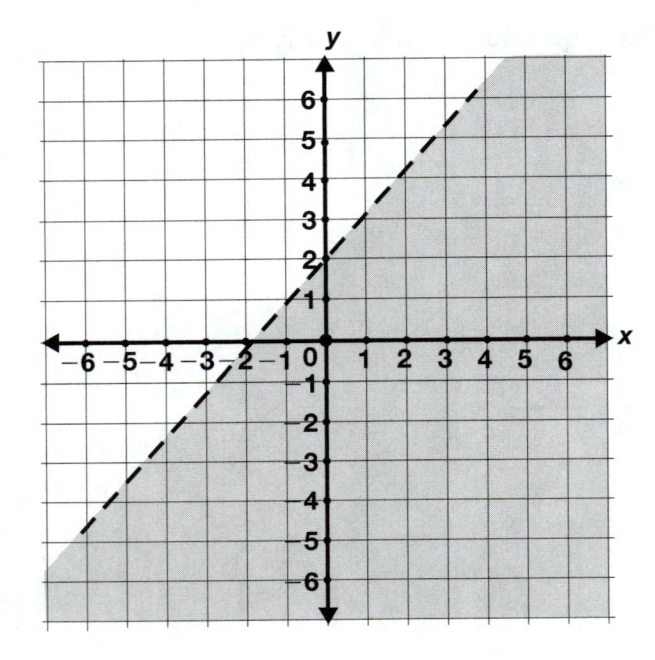

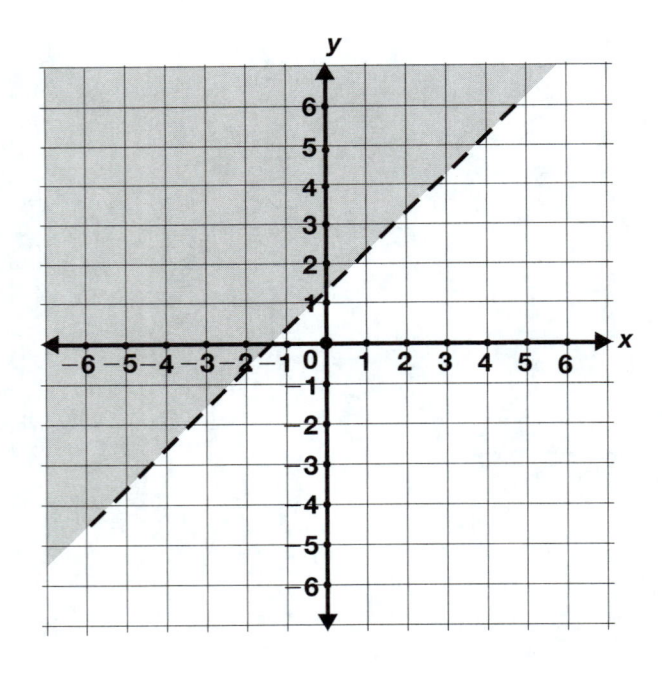

H

J

3 The initial pressure inside a closed container is 100 pounds per square inch (psi). As the temperature inside the container decreases, the pressure decreases. If the pressure decreases 1.9 psi for each degree Fahrenheit of decreased temperature, which equation best represents *p*, the pressure inside the container after the temperature has decreased *t* degrees?

A $p = \dfrac{\left(\frac{1}{9}t\right)}{100}$

B $p = 100t - 1.9$

C $p = 100 - 1.9t$

D $p = 1.9t$

4 A scientist released a study that the more it rains, the higher trees grow. What is the independent quantity in this relationship?

F The amount of rain

G The height of trees

H Both data

J Cannot be determined

Chapter 3 Answer Explanations

Let's Review 3

1. B
The number of party favors ordered is dependent upon the number of guests at the party, so answer choice B is correct.

2. G
The second graph crosses the y-axis and other vertical lines two times, so it is not a function.

3. 10, 12
For this function, you need to multiply 5 and 6 by 2.

Let's Review 4

1. B
Since we don't know Melanie's age, but we know that Melanie's age is 2 + Rachel's age, $x = y + 2$. Solving for Rachel's age (y) in terms of Melanie's age (x) yields answer choice B.

2. J
If you plug values into the inequality, you'll see that answer choice J is correct. Remember that the line for $y = x - 1$ should be dashed, and the side of the line that contains values that make the inequality true should be shaded.

3. B
By substituting the values $x = 6$ and $y = 12$ into the answers, you can see that only answer choice B is true.

4. G

If you leased the car for one week and drove it 500 miles, it would cost $1,250. Answer choice G is the correct answer.

Chapter 3 Practice Problems

1. C

At first the bus is traveling at its maximum speed, so the line should be straight and high on the graph. Then the bus slows down, which means that the line will drop, and then speeds up again, which means the line should go up. Note that changes in speed are not instantaneous, so the lines as the bus slows down and speeds up are not vertical, but slanted. Answer choice C depicts this situation.

2. G

Substitute numbers for x into $y = x + 2$, and plot (x, y) on the graph, making this line solid. Then check a point below the line, such as $(0, 0)$ to see whether it makes the inequality true. The second graph is the only one with the correct line (solid) and correct area shaded.

3. C

If the initial pressure is 100 and it is decreasing, you will have to subtract. The problem doesn't say how many degrees the temperature drops, so t is variable, and 1.9 the amount the pressure decreases per degree, would be multiplied by this variable.

4. F

The independent variable would be the one that causes the dependent variable to change. The amount of rain causes the trees to grow, so the amount of rain is the independent variable.

Chapter 4

Functions, Part 2

$f(x)=x^3$
$2x+1y=12$
$y=x^3-4$

Objectives

Objective 2: The student will demonstrate an understanding of the properties and attributes of functions.

A(b)(2) **Foundations for functions.** The student uses the properties and attributes of functions.

(A) The student identifies [and sketches] the general forms of linear ($y = x$) and quadratic ($y = x^2$) parent functions.

(B) For a variety of situations, the student identifies the mathematical domains and ranges and determines reasonable domain and range values for given situations.

(D) In solving problems, the student [collects and] organizes data, [makes and] interprets scatterplots, and models, predicts, and makes decisions and critical judgments.

Objective 3: The student will demonstrate an understanding of linear functions.

A(c)(1) **Linear functions.** The student understands that linear functions can be represented in different ways and translates among their various representations.

(A) The student determines whether or not given situations can be represented by linear functions.

(C) The student translates among and uses algebraic tabular, graphical, or verbal descriptions of linear functions.

A(c)(2) **Linear functions.** The student understands the meaning of the slope and intercepts of linear functions and interprets and describes the effects of changes in parameters of linear functions in real-world and mathematical situations.

(A) The student develops the concepts of slope as a rate of change and determines slopes from graphs, tables, and algebraic expressions.

(B) The student interprets the meaning of slope and intercepts in situations using data, symbolic representations, or graphs.

(C) The student investigates, describes, and predicts the effects of changes in m and b on the graph of $y = mx + b$.

(D) The student graphs and writes equations of lines given characteristics, such as two points, a point and a slope, or a slope and y-intercept.

(E) The student determines the intercepts of linear functions from graphs, tables, and algebraic representations.

(F) The student interprets and predicts the effects of changing slope and y-intercept in applied situations.

(G) The student relates direct variation to linear functions and solves problems involving proportional change.

Objective 5: The student will demonstrate an understanding of quadratic and other nonlinear functions.

A(d)(1) **Quadratic and other nonlinear functions.** The student understands that the graphs of quadratic functions are affected by the parameters of the function and can interpret and describe the effects of changes in the parameters of quadratic functions.

(B) The student investigates, describes, and predicts the effects of changes in a on the graph of $y = ax^2$.

(C) The student investigates, describes, and predicts the effects of changes in c on the graph of $y = x^2 + c$.

(D) For problem situations, the student analyzes graphs of quadratic functions and draws conclusions.

A(d)(2) Quadratic and other nonlinear functions. The student understands there is more than one way to solve a quadratic equation and solves them using appropriate methods.

(A) The student solves quadratic equations using [concrete] models, tables, graphs, and algebraic methods.

(B) The student relates the solutions of quadratic equations to the roots of their functions.

A(d)(3) Quadratic and other nonlinear functions. The student understands there are situations modeled by functions that are neither linear nor quadratic and models the situations.

(A) The student uses [patterns to generate] the laws of exponents and applies then in problem-solving situations.

Linear Functions

A **linear function** is a function that forms a straight line when graphed. In a linear function, x is not raised to a higher power than 1. A linear function can be written in the form of

$$f(x) = ax + b$$

where a and b are real numbers.

The following are some examples of linear functions:

$$y = 2x - 1$$

$$y = 3x + 2$$

Consider this simple linear function: $y = x + 2$

If you listed the numbers that could be substituted for the variables x and y, your list might look like this:

x	y
1	3
2	4
3	5
4	6

If you graphed these coordinates and drew a line, it would look like this:

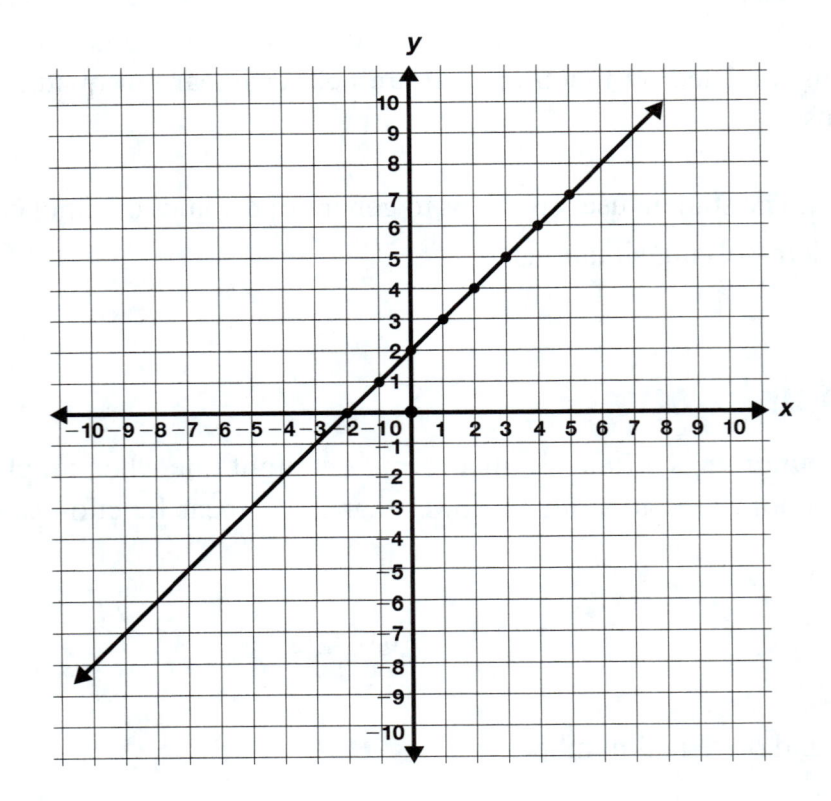

Chapter 4: Functions, Part 2

Which graph represents a linear function?

A

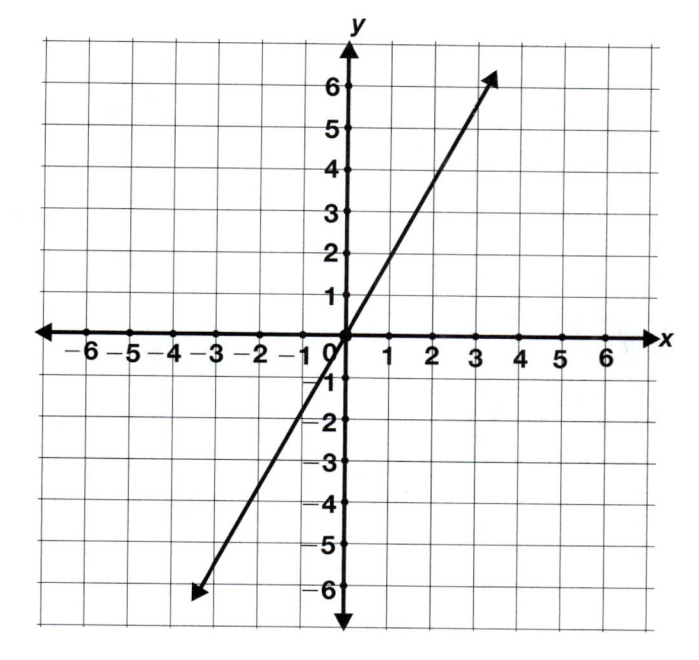

B

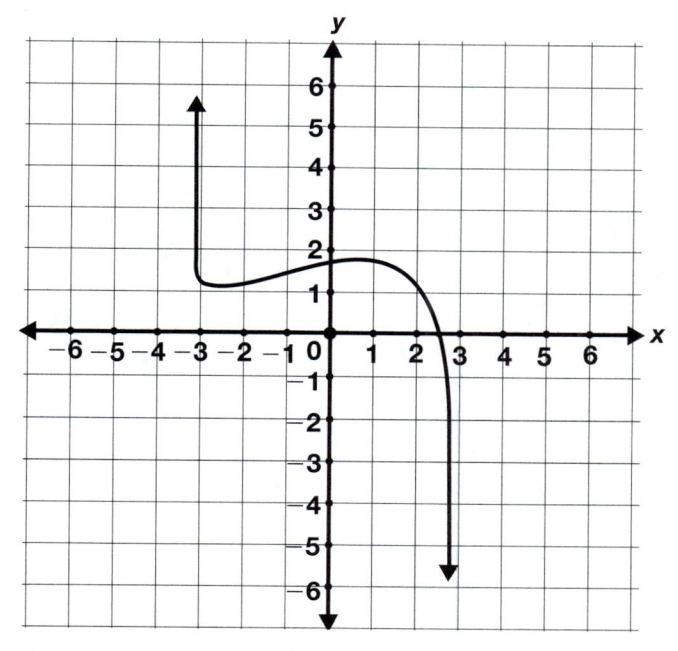

86 TAKS
Mathematics

C

D

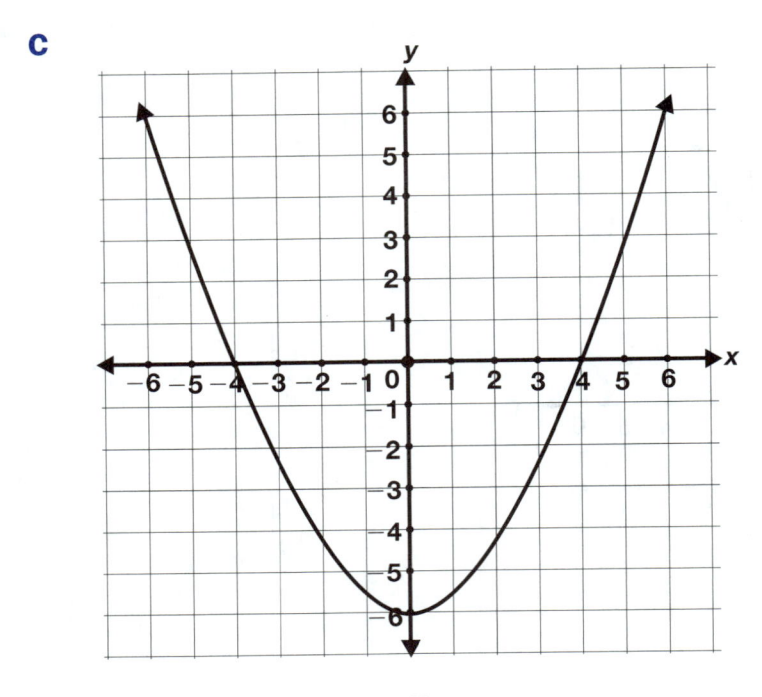

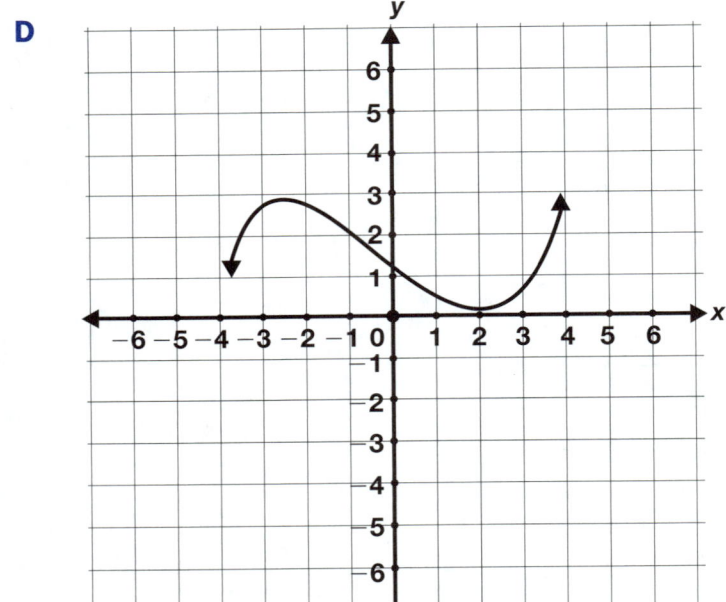

Remember that a linear function is a straight line, so answer choice A is correct.

Quadratic Functions

Quadratic functions are functions in the form of

$$f(x) = ax^2 + bx + c$$

where *a*, *b*, and *c* are real numbers and *a* is not equal to zero. The following are quadratic functions:

$$f(x) = x^2 + 3x + 2$$

$$f(x) = x + 2x^2 - 1$$

When graphed, quadratic functions produce a **parabola**, a curve that looks like the letter U or an upside-down U. The **vertex** of a parabola is the point at which it changes direction. Parabolas can vary in width, but they all have the same basic shape. The curve in this graph is a parabola:

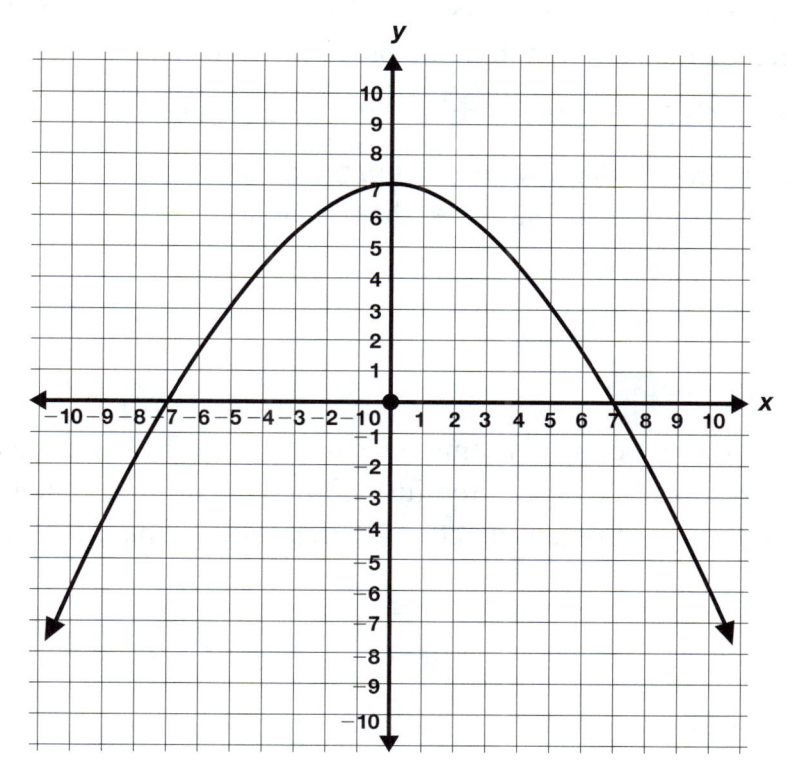

The path of a ball when it is thrown forms a parabola. The parabola begins when the ball is thrown. It reaches its highest point, its vertex, right before gravity starts to pull it downward. If the ball continues to bounce after it hits the ground, other parabolas are formed.

Note that if a parabola opens downward, the vertex is the highest point. Conversely, if it opens upward, the vertex is the lowest point.

You might be asked a question about changing the width of a parabola or moving a parabola up or down on the coordinate grid. The following are some general rules about the shape and movement of parabolas:

- Positive values of a, the coefficient of x^2, cause the parabola to open upward.
- Negative values of a, the coefficient of x^2, cause the parabola to open downward.
- A negative c in the equation $f(x) = ax^2 + bx + c$ causes some or all points on the parabola to lie below the x-axis.
- As c increases, the vertex of the parabola gets higher.
- The greater the absolute value of a in the equation $f(x) = ax^2 + bx + c$, the narrower the parabola.

Parent Functions

The phrase "parent function" is sometimes used on the TAKS. It's a basic function on which other functions are based. You just learned about linear and quadratic functions; they are two kinds of parent functions. Other kinds of parent functions include:

- Absolute function – A function with an absolute value, such as $f(x) = |x - 2|$. An absolute value function looks like a V when it is graphed. If you have forgotten what absolute value means, refer back to Chapter 1. The graph of $f(x) = |x|$ is shown below.

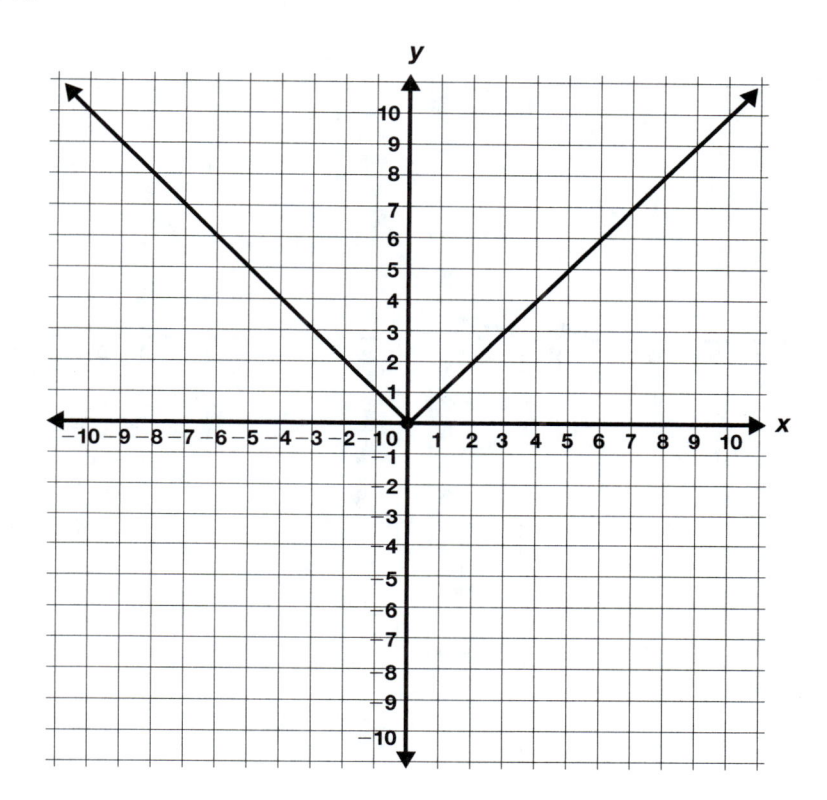

- Exponential function – A function in which the variable is an exponent, such as $f(x) = a \cdot b^x$. An exponential function has the general shape shown below when it is graphed.

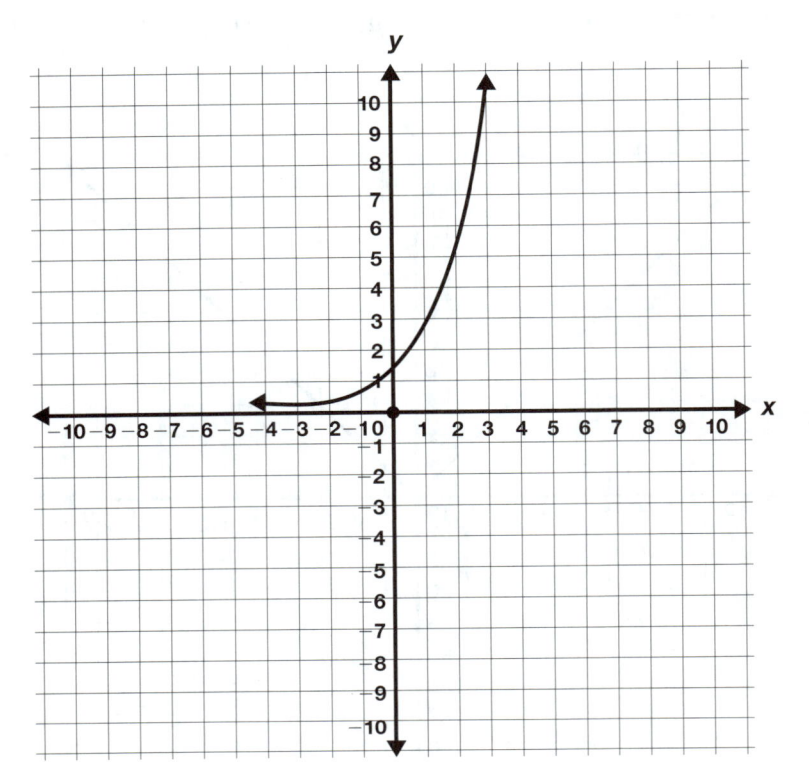

Chapter 4: Functions, Part 2

- Cubic function—a function in which the independent variable is raised to the third power, such as $f(x) = ax^3 + bx^2 + cx + d$, where $a \neq 0$ (b, c, and d can equal 0). It generally looks like the shape shown below when it is graphed.

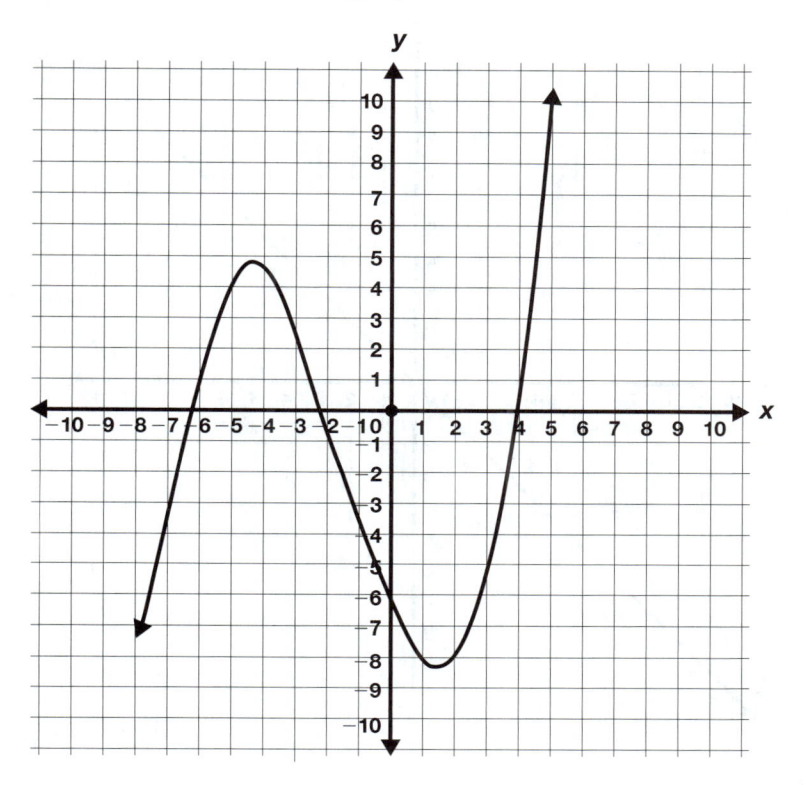

Read this question:

The graph below represents which type of parent function?

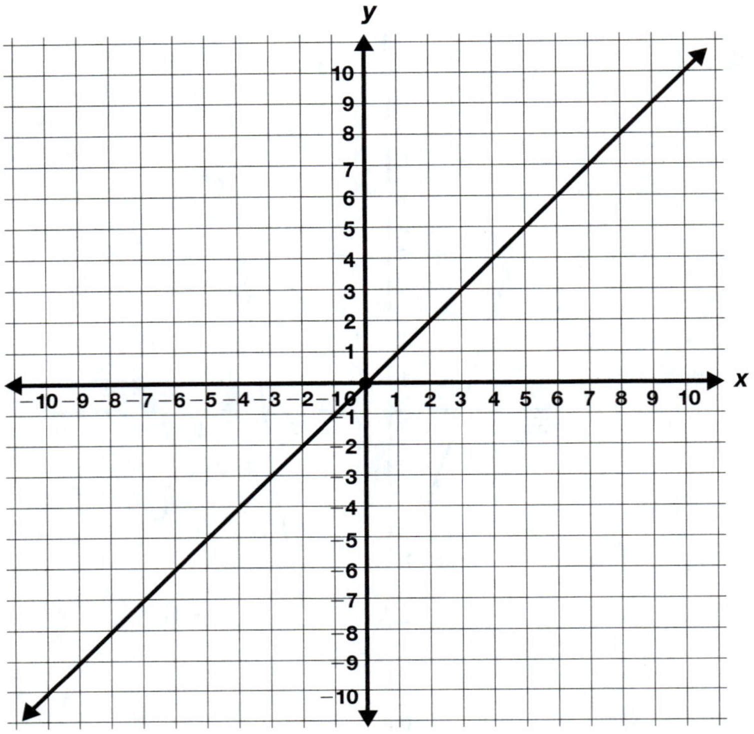

F Quadratic

G Absolute value

H Exponential

J Linear

From what you have just learned, you should be able to tell that this is the graph of a linear function—it's straight!

Domain and Range

You learned earlier that you can define a set of coordinates for a function. The **domain** is the list of x-coordinates, and the **range** is the list of y-coordinates. For example,

domain: {−3, −2, −1, 0, 1, 2}
range: {5}
represents the coordinates (−3, 5), (−2, 5), (−1, 5), (0, 5), (1, 5), and (2, 5).

Even though all of the *y*-coordinates are 5, each pair of coordinates represents a different point on a coordinate plane. In this case, the line passing through the points is parallel to the *x*-axis. Therefore, this domain and range represent a function.

If the *x*-coordinates (values in the domain) repeat themselves, however, the points will not represent a function because they will fail the vertical line test.

x- and y-Intercepts

x-intercepts are the points at which a graph crosses the *x*-axis. **y-intercepts** are the points at which a graph crosses the *y*-axis. Since $y = 0$ all along the *x*-axis, to find the *x*-intercept in an equation, substitute 0 for *y* and solve the equation.

$$2x + y = 12$$
$$2x + 0 = 12$$
$$\frac{2x}{2} = \frac{12}{2}$$
$$x = 6$$

The *x*-intercept is the point (6, 0).

Follow a similar procedure to find the *y*-intercept by substituting 0 for *x*:

$$2x + y = 12$$
$$2(0) + y = 12$$
$$0 + y = 12$$
$$y = 12$$

The *y*-intercept is the point (0, 12).

Slope

The **slope** of a line indicates that line's steepness; the greater the slope, the more steep the line. To determine the slope of a line, use this formula, which is called the **rise-over-run formula**:

$$\text{Slope} = \frac{y_2 - y_1}{x_2 - x_1}$$

where the slope is calculated from any two points on the line: (x_1, y_1) and (x_2, y_2).

The slope of a line can be positive, negative, zero, or undefined. Lines with a positive slope slant upward from left to right. Lines with a negative slope slant downward from left to right. Lines with zero slope are horizontal lines (they have zero "rise," meaning all of the y coordinates are the same). Lines with an undefined slope are vertical lines (all points have the same x coordinate, so the rise-over-run formula has a zero denominator, which is called undefined because you cannot divide by 0).

As an example of using the rise-over-run formula, suppose a line has two points with coordinates (1, 5) and (4, −3). You would use the formula this way:

$$\text{Slope} = \frac{(y_2 - y_1)}{(x_2 - x_1)} = \frac{(-3 - 5)}{(4 - 1)} = -\frac{8}{3}$$

The slope of this line is therefore $-\frac{8}{3}$.

Slope-Intercept Form of a Linear Function

As discussed in Chapter 4, the **y-intercept** of a linear function is the point at which the line crosses the y-axis; $x = 0$ at the y-intercept. Any linear function can be expressed in the slope-intercept form. You can tell the slope and y-intercept of a linear function just by looking at the equation in this form:

$$y = mx + b$$

where m is the slope of the line, and b is the y-intercept.

Let's look at a problem to see how this form of a linear equation is useful.

What is the slope of the line $2y = 4x - 7$?

One way to answer this is to find two points, (x_1, y_1) and (x_2, y_2), that are on the line and to use the rise-over-run formula to calculate the slope. But an easier way to find the slope for this function is to put it in slope-intercept form by dividing both sides of the equation by 2:

$$\frac{2y}{2} = \frac{4x}{2} - \frac{7}{2}$$

$$y = 2x - \frac{7}{2}$$

You can easily see that the slope is 2. You can also see that the *y*-intercept of this line is $-\frac{7}{2}$.

Let's Review 5: Slope

Complete each of the following questions. Use the Tip below each question to help you choose the correct answer. When you finish, check your answers with those at the end of Chapter 4.

1 Which of the following describes the slope of a line parallel to the x-axis?

A   Positive slope

B   Negative slope

C   Zero slope

D   Undefined slope

TIP
If you don't remember what you learned about slope, reread this section to find the answer.

2 What is the slope of a line that passes through the points (2, 5) and (6, 13)?

F   −2

G   0

H   1

J   2

TIP
Use the rise-over-run formula to find the slope of this line.

Chapter 4: Functions, Part 2   97

3 The following figure shows the net for a three-dimensional object.

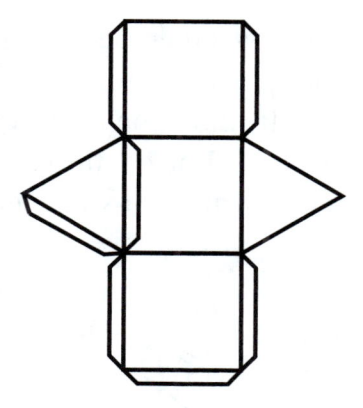

When folded, which object will this net produce?

A

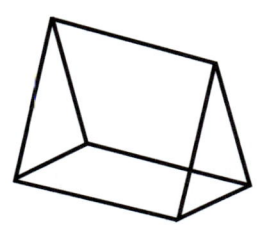

B

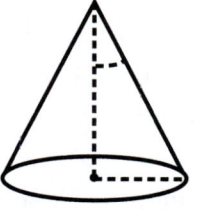

C

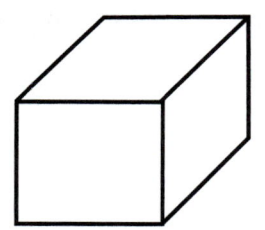

D

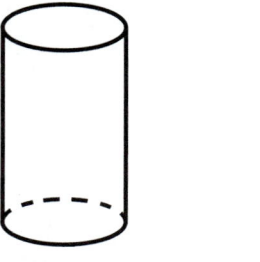

TIP
Look carefully at the figure. What would you get if you folded the net on the dotted lines?

Scatterplots

A **scatterplot** shows at a glance whether there is a relationship between two sets of data. In a scatterplot, data are plotted by using dots. If the dots show a trend, a line or curve is drawn that approximates the direction of the data. If the line goes up from left to right, the trend is said to be positive. Look at the scatterplot shown here. It shows the relationship between the hours of practice for students who play a musical instrument and their test scores.

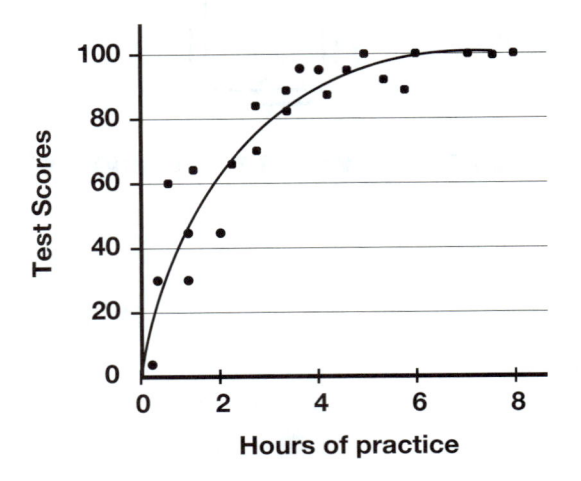

According to this scatterplot, students who play a musical instrument do have higher test scores. There is a positive trend, as shown by the curve. In fact, the more hours per week they practice playing an instrument, the higher their test scores are.

Now look at the next scatterplot. It compares the daily hours of watching TV and students' test scores. This time the scatterplot shows a negative trend, as seen by the line.

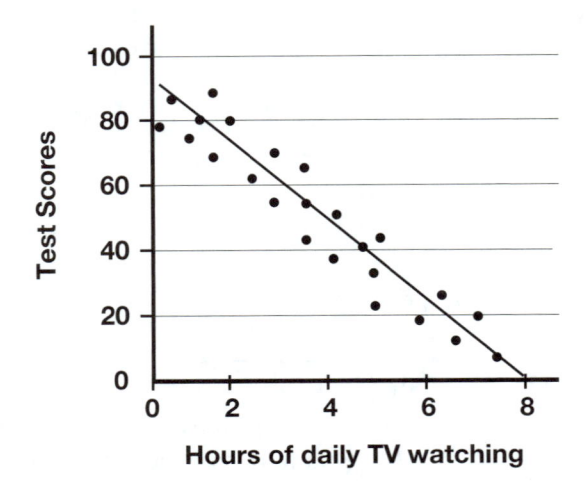

This scatterplot shows that students who watch more TV get lower grades. The more time they spend watching TV, the lower are their test scores.

Sometimes a scatterplot will show no trend between two variables. The scatter plot below shows no trend between hours of sleep per night and getting good test scores.

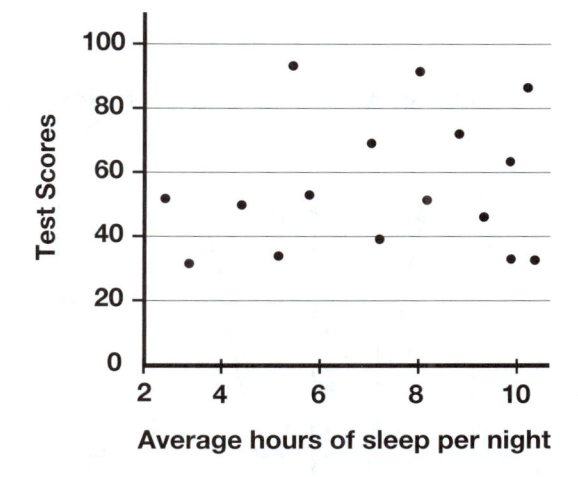

Let's Review 6: Graphs and Equations

Complete each of the following questions. Use the Tip below each question to help you choose the correct answer. When you finish, check your answers with those at the end of Chapter 4.

1 The graph of the equation $y = x^2$ is shown below.

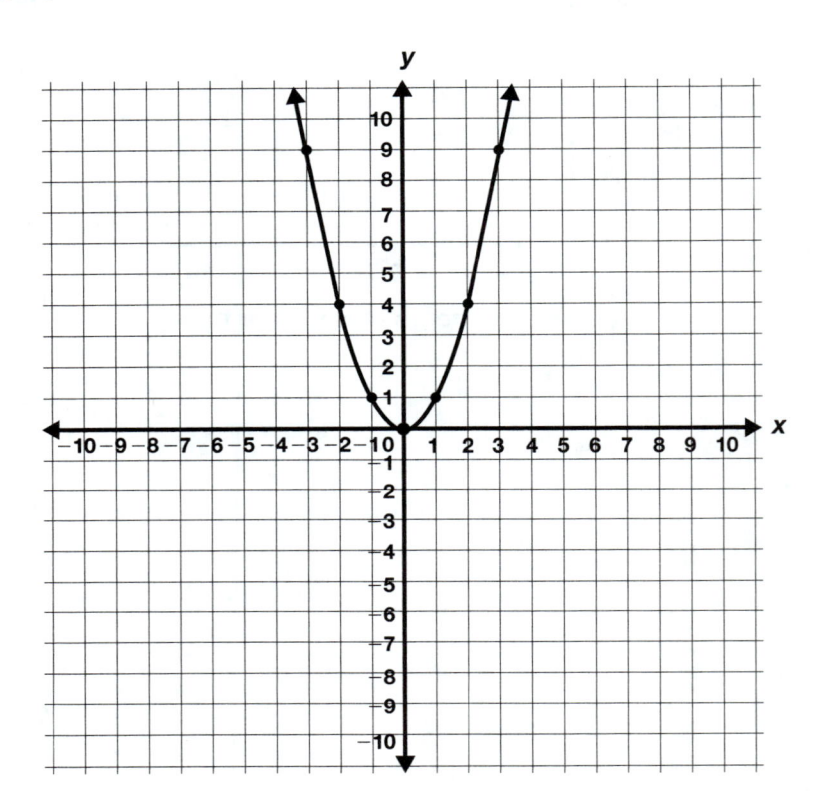

If the graph is translated 3 units up, what will be the equation of the resulting graph?

A $y = 3x^2$

B $y = 3 + x^2$

C $y = \dfrac{x^2}{3}$

D $y = x^2 - 3$

TIP You're moving up 3 units on the *y*-axis.

2 Which of the following is a sufficient condition to show that a certain equation does not represent a quadratic equation?

F The graph has no x-intercepts.

G The graph looks like the letter U.

H The graph is a straight line.

J The graph of the equation has more than one x-intercept.

TIP: Choose the answer that could not possibly be a quadratic equation.

3 The graph below represents which type of parent function?

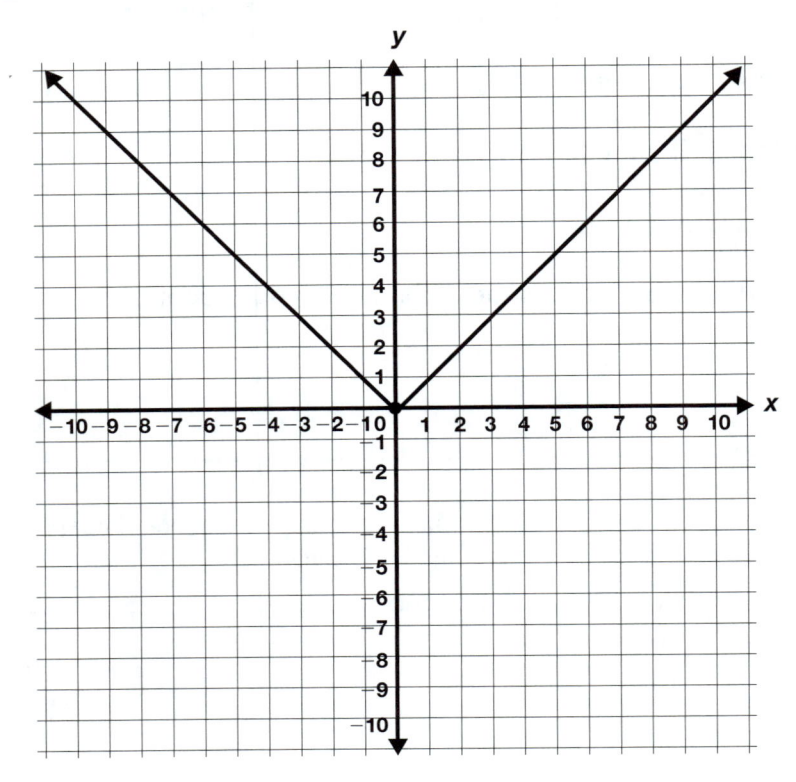

- **A** Exponential
- **B** Linear
- **C** Quadratic
- **D** Absolute value

TIP
Choose the parent function that, when graphed, looks like the letter V.

4 Which of the following polynomial equations best represents this graph?

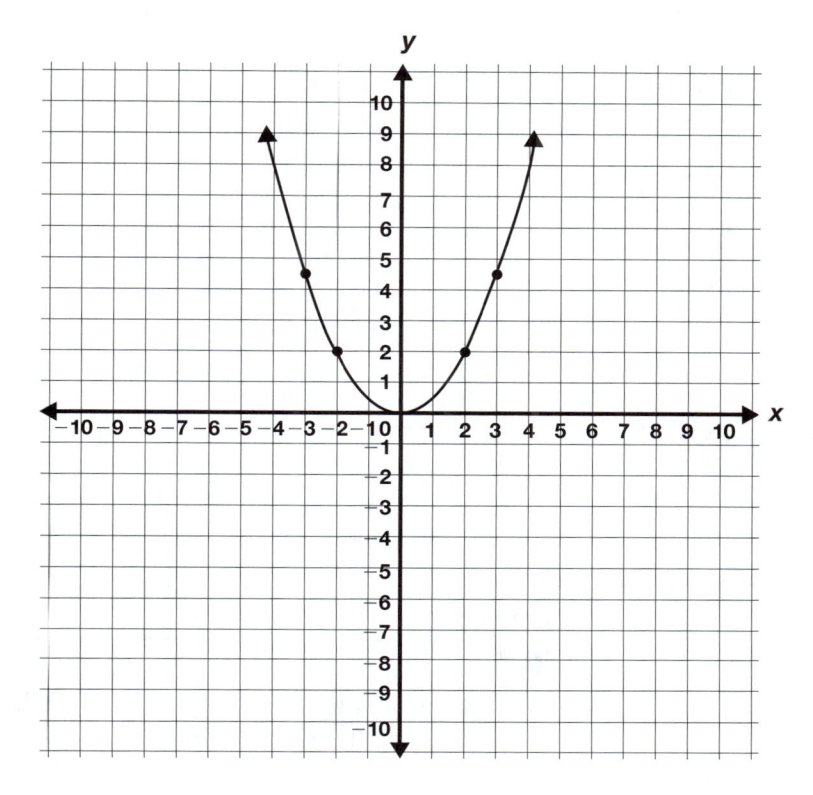

F $y = 2x^2$

G $y = \dfrac{1}{4}x^2$

H $y = \dfrac{1}{2}x^2$

J $y = x^2$

TIP
Find some coordinates for each equation. Choose the equation that matches the graph.

5 Find the *x*- and *y*-intercepts of 3*x* + 2*y* = 24.

A   *x*-intercept   (0, 8)

       y-intercept   (12, 0)

B   *x*-intercept   (8, 1)

       y-intercept   (1, 12)

C   *x*-intercept   (1, 8)

       y-intercept   (12, 1)

D   *x*-intercept   (8, 0)

       y-intercept   (0, 12)

TIP

Remember to substitute $y = 0$ to find the *x*-intercept and $x = 0$ to find the *y*-intercept.

Chapter 4 Practice Problems

Complete each of the following practice problems. Check your answers at the end of this chapter. Be sure to read the answer explanations!

1 Which of the following is the vertex of the graph of the equation $4y = x^2 - 6x + 1$?

A (3, 2)

B (2, 3)

C (3, −2)

D (−3, 2)

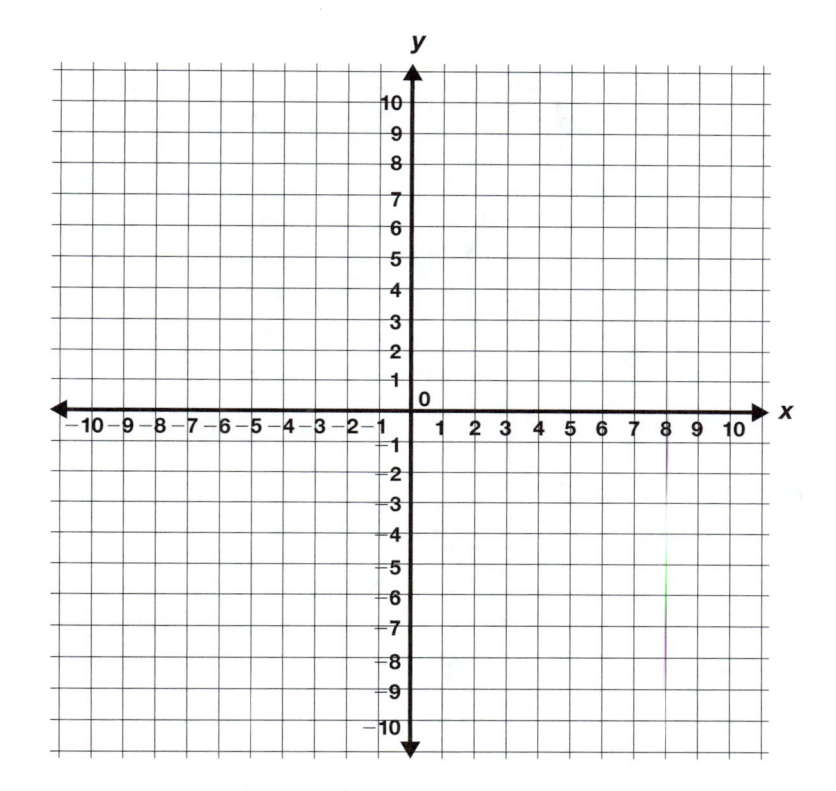

2 What is the effect on the graph of the equation $y = -3x^2$ when the equation is changed to $y = 3x^2$?

F The x values for any given y are farther from the y-axis.

G The graph is rotated 90° about the origin.

H The graph of $y = -3x^2$ is a reflection of $y = 3x^2$ across the x-axis.

J The x values for any given y are closer to the y-axis.

3 In which of the following parabolas does the coefficient *a* in $f(x) = ax^2 + bx + c$ have the greatest value?

A

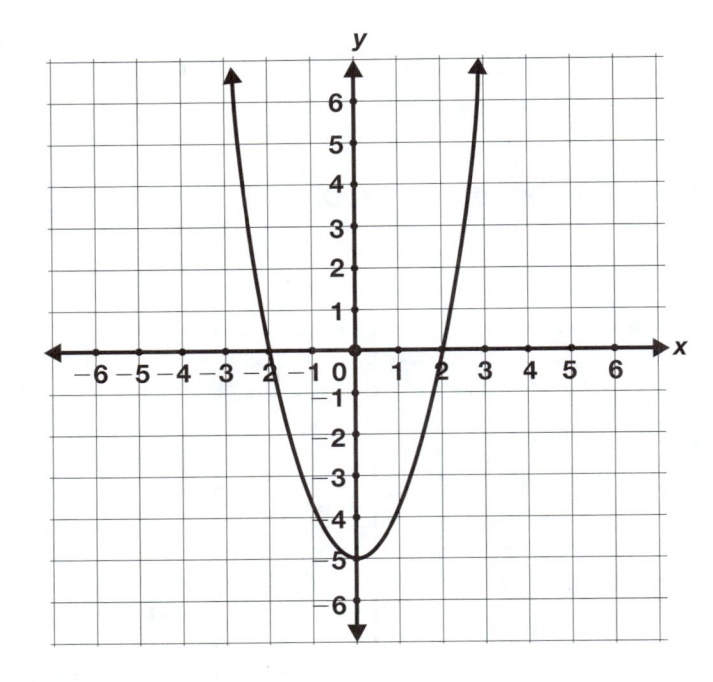

B

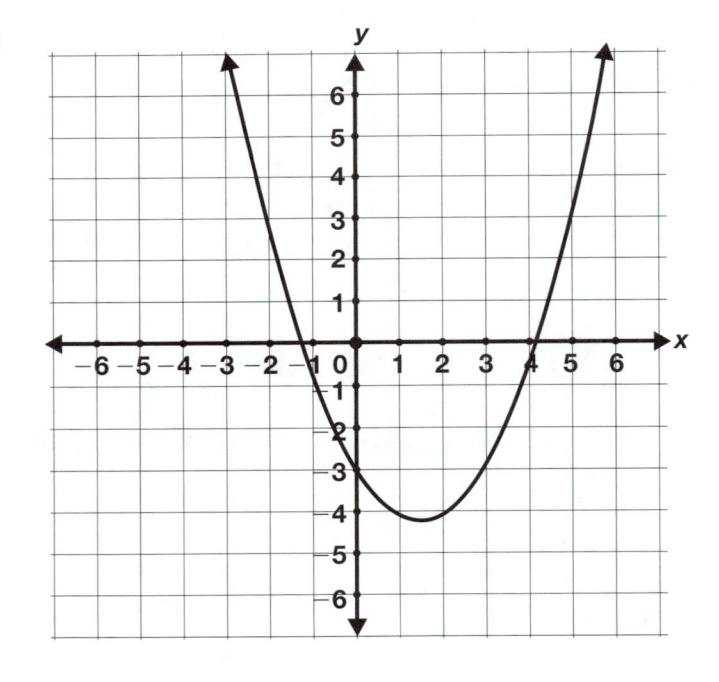

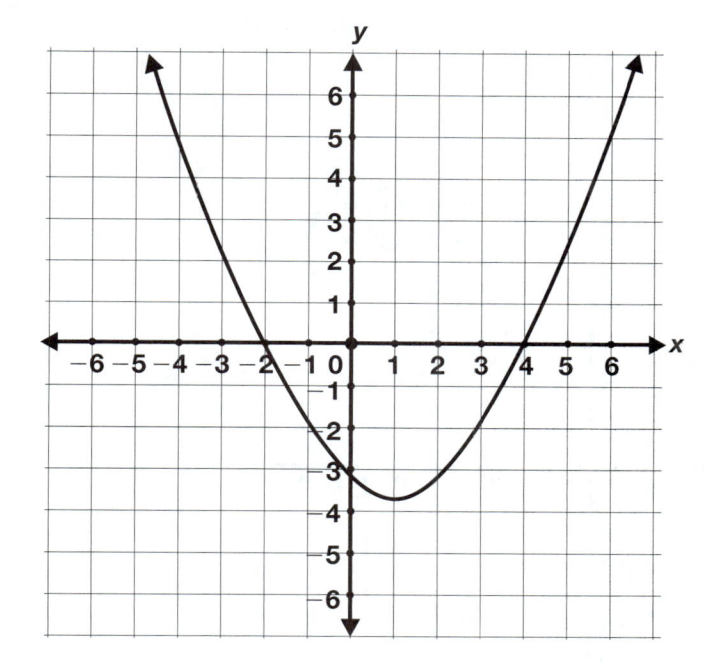

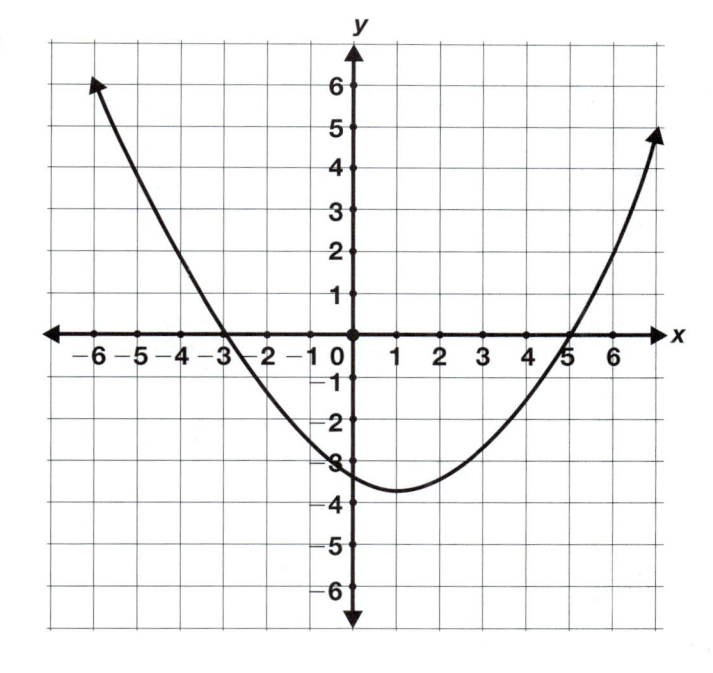

C

D

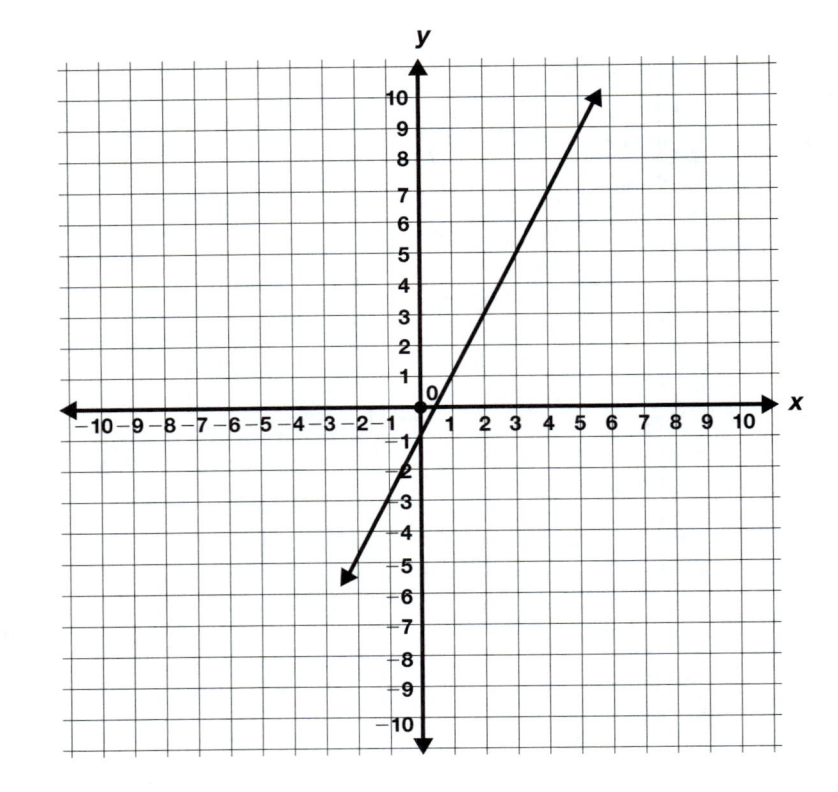

4. What is the slope of the line in the graph shown below?

F 0

G 4

H 2

J 1

Chapter 4: Functions, Part 2

Chapter 4 Answer Explanations

Let's Review 5: Slope

1. C

All lines parallel to the *x*-axis have a zero slope.

2. J

The slope may be found by calculating $\frac{(13 - 5)}{(6 - 2)} = \frac{8}{4} = 2$.

3. A

Let's Review 6: Graphs and Equations

1. B

In $y = ax^2 + bx + c$, you need to increase the constant *c* by 3 to move this parabola three units up the *y*-axis.

2. H

A quadratic equation is not a straight line, so answer choice H is correct.

3. D

A parent function that is an absolute value function looks like a V.

4. H

If you substitute some values into each equation, you'll see that answer choice H is the correct answer.

5. D

You need to substitute 0 for *y* to find the *x*-intercept, and 0 for *x* to find the *y*-intercept. If you do this, you'll see that the *x*-intercept is (8, 0) and the *y*-intercept is (0,12).

Chapter 4 Practice Problems

1. C

Substitute some values into the equation and plot the parabola. When you do you'll see that the lowest point, the vertex, is (3, –2).

2. H

The negative sign means the graph opens downward. If it's changed to positive, the graph opens upward, and is reflected across the *x*-axis.

3. A

The greater the absolute value of the coefficient *a* in the equation $f(x) = ax^2 + bx + c$ the narrower the parabola. Therefore, answer choice A is correct.

4. H

Begin by listing two sets of coordinates for the line, such as (2, 3), and (3, 5). Then use the rise-over-run formula to find the slope: $\frac{(y_2 - y_1)}{(x_2 - x_1)} = \frac{(5 - 3)}{(3 - 2)} = \frac{2}{1} = 2$

Chapter 5

Geometry and Spatial Sense, Part 1

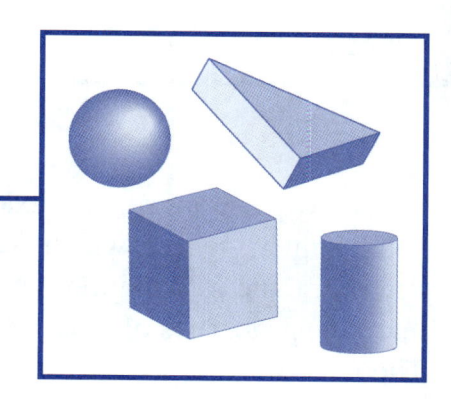

Objectives

Objective 6: The student will demonstrate an understanding of geometric relationships and spatial reasoning.

G(b)(4) **Geometric structures.** The student uses a variety of representations to describe geometric relationships and solve problems.

(A) The student selects an appropriate representation ([concrete], pictorial, graphical, verbal, or symbolic) in order to solve problems.

G(c)(1) **Geometric patterns.** The student identifies, analyzes, and describes patterns that emerge from two- and three-dimensional geometric figures.

(A) The student uses numeric and geometric patterns to make generalizations about geometric properties, including properties of polygons, ratios in similar figures and solids, and angle relationships in polygons and circles.

(B) The student uses the properties of transformations and their compositions to make connections between mathematics and the real world in applications such as tessellations or fractals.

(C) The student identifies and applies patterns from right triangles to solve problems, including special right triangles (45°-45°-90° and 30°-60°-90°) and triangles whose sides are Pythagorean triples.

Some questions on the Texas Assessment of Knowledge and Skills (TAKS) will be about similar figures, which are figures with the same proportions. You'll learn how to determine the missing measure of a side in similar figures in this chapter. You'll also learn the properties of lines and angles in this chapter and how to use the Pythagorean Theorem.

Plane Figures

A **plane figure** is a flat, closed figure. The following table lists plane figures and their properties.

PLANE FIGURES

Figure	Description	Example
Parallelogram	A parallelogram is a **quadrilateral** (four-sided figure) with opposite sides that are parallel and equal in length. The opposite angles are equal and any two consecutive angles are supplementary. The sum of the angles in a parallelogram is 360°. Each diagonal of a parallelogram divides it into two triangles that are the same size and shape.	
Rectangle	A rectangle is a special type of parallelogram, which means it has all of the properties of a parallelogram, plus it has four right angles.	
Square	A square is another special type of parallelogram. It has four right angles and all four sides are equal in length.	
Rhombus	A rhombus is yet another special type of parallelogram. It has four equal sides, but may or may not have right angles. If a parallelogram has congruent adjacent sides, it's a rhombus. (So a square is actually a rhombus with all right angles.)	

(Continued on next page)

PLANE FIGURES (Continued)

Figure	Description	Example
Trapezoid	A trapezoid is a quadrilateral, but it is not a parallelogram. It has only one pair of parallel sides, and may or may not have a right angle. The sum of the angles in a trapezoid is 360°. In an isosceles trapezoid, the base angles are congruent.	
Triangle	A triangle has three sides, the length of sides can vary, and it may or may not have a right angle. The sum of the angles in a triangle is 180°.	
Circle	A circle has no sides. All the points on a circle are the same distance from the center. There are 360° in a circle.	
Pentagon	A pentagon has five sides that may or may not be equal; it can have up to three right angles. The sum of the angles in a pentagon is 540°. A regular pentagon has equal sides and each angle measures 108°.	
Hexagon	A hexagon has six sides that may or may not be equal; it can have up to three right angles. The sum of the angles in a hexagon is 720°. A regular hexagon has equal sides and each angle measures 120°.	
Octagon	An octagon has eight sides that may or not be equal; it can have up to three right angles. The sum of the angles in an octagon is 1,080°. A regular octagon has equal sides and each angle measures 135°.	

Congruent Figures

Figures that are **congruent** are exactly the same size and shape. If you place two congruent figures on top of each other, they will fit exactly. The rectangles below are congruent.

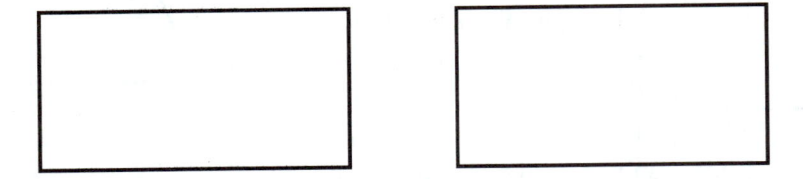

These triangles are also congruent, even though they appear to be facing different directions.

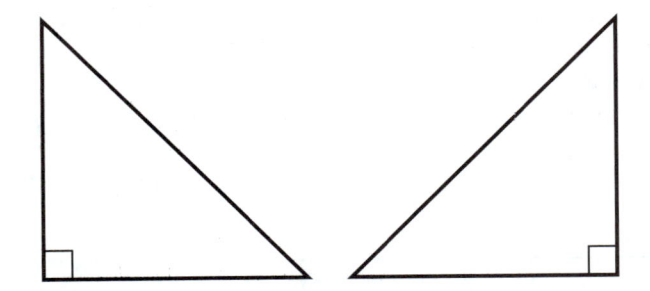

Similar Figures

Figures that are similar are not necessarily congruent. They have the same shape but they aren't necessarily the same size. If figures are similar, their corresponding sides can be written as a proportion because one figure is an enlargement of the other. Their corresponding angles are equal.

These triangles are similar:

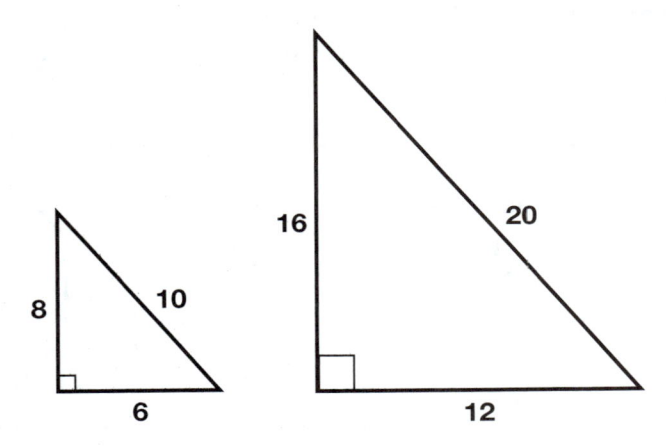

Chapter 5: Geometry and Spatial Sense, Part 1 — 115

These rectangles are similar:

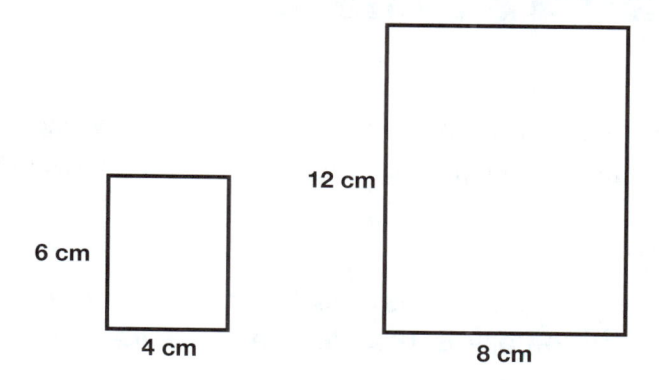

Note that the corresponding sides of these two similar rectangles can be written as a proportion that is the same when it is reduced:

$$4 : 8 = 1 : 2$$
$$6 : 12 = 1 : 2$$

By using the fact that corresponding sides of similar figures are proportional, you can determine a missing side for similar shapes. For example, read this question:

Find the missing length (x) for the pair of similar figures below.

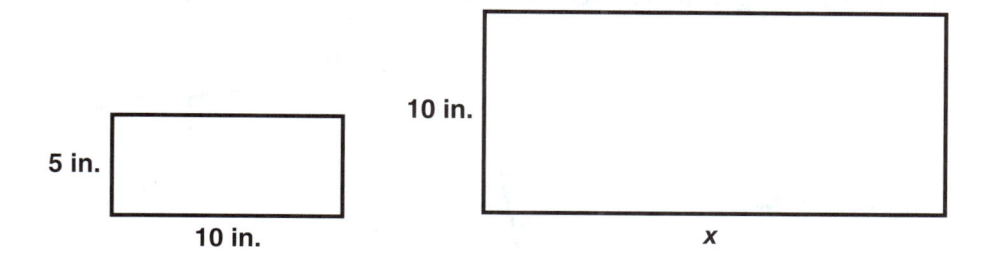

Because similar figures have proportional corresponding sides, the proportion for the short sides must be the same for the long sides of these rectangles.

$$5 : 10 = 1 : 2$$
$$10 : x = 1 : 2$$
$$x = 20 \text{ inches}$$

$$\frac{5}{10} \stackrel{2}{\wedge} \frac{10}{10 \cdot 2 = 20} \quad 20 \text{ inches.}$$

Let's Review 7: Plane Figures

Complete each of the following questions. Use the Tip below each question to help you choose the correct answer. When you finish, check your answers with those at the end of Chapter 5.

1 Jamie has a picture that measures 12 inches in width and 24 inches in length. If Jamie enlarges the picture to make a poster 2 feet wide, how long will the poster be?

- A 2 ft
- B 4 ft
- C 6 ft
- D 8 ft

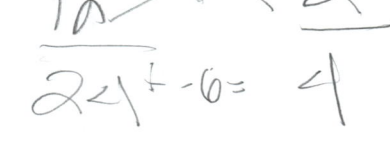

TIP: Two feet equals 24 inches, and the proportion for 12 : 24 is 1 : 2.

2 Find the missing length (*x*) for the pair of similar figures below.

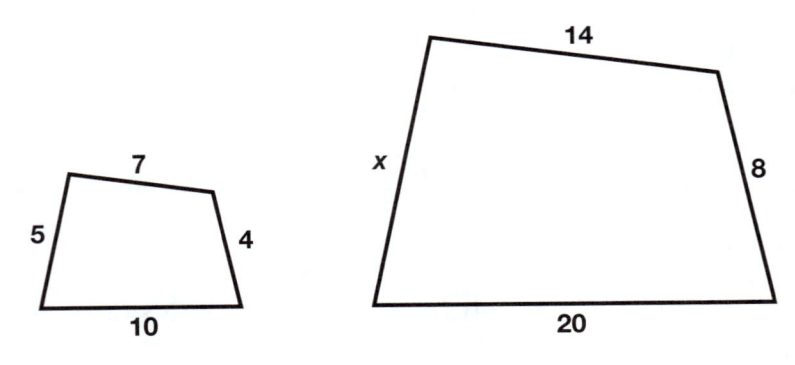

- F 6
- G 8
- H 10
- J 20

TIP: The corresponding side on the smaller figure is 5.

Chapter 5: Geometry and Spatial Sense, Part 1   117

3 If the two triangles here are similar, find the measure of side *x*.

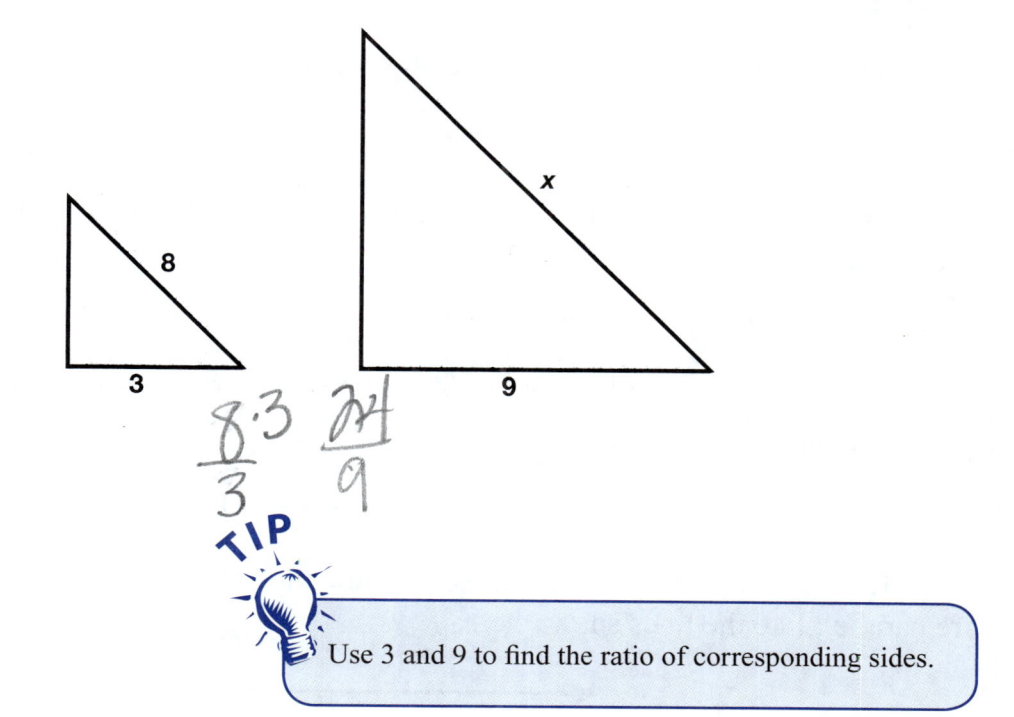

A 12
B 16
C 24
D 48

TIP Use 3 and 9 to find the ratio of corresponding sides.

4 Which of the following has four right angles?

F a rhombus
G a trapezoid
H an octagon
J a rectangle

TIP If you're not sure of this answer, go back and read the properties of plane figures.

Lines

A **line** has an arrow on both ends to show that it keeps going.

A **line segment** is part of a line. It has two **endpoints**, to show that it stops and doesn't keep going.

A **ray** is also part of a line, but unlike a line segment, it keeps going in one direction. A ray has only one endpoint.

Lines that never intersect are called **parallel lines**. For example, strings on a guitar are parallel, like the lines shown here:

Lines that intersect to form 90° angles (right angles) are called **perpendicular lines**. The place where the lines intersect is called the **point of intersection**.

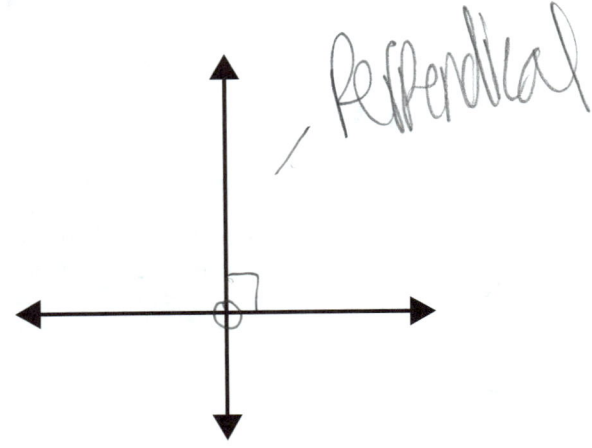

Angles

Two rays with a common endpoint form an **angle**. The place where they join is called the **vertex**.

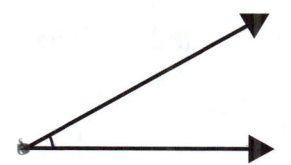

Angles can be classified by their size. The following table shows the names of angles according to the number of degrees they contain.

ANGLES

Angle	Description	Example
Acute angle	Less than 90 degrees.	
Right angle	Exactly 90 degrees.	
Obtuse angle	Greater than 90 degrees and less than 180 degrees.	
Straight angle	Exactly 180 degrees.	
Reflex angle	Greater than 180 degrees.	

Angle Relationships

The sign ∠ stands for the word "angle." **Adjacent angles** are angles that share a side.

In the illustration below, ∠ADC and ∠CDB are adjacent angles. They share ray $\overrightarrow{DC}$.

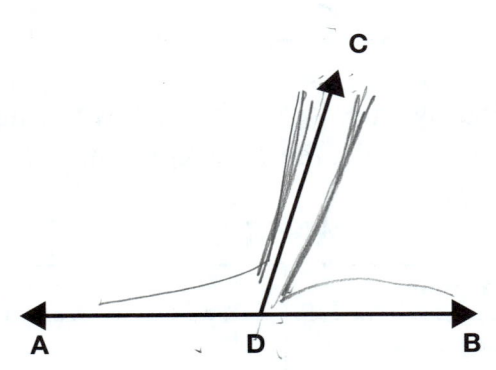

Two angles that add up to 90° are called **complementary angles.** Two angles that form a straight angle or add up to 180° are called **supplementary angles.** The angles above are supplementary because they form a straight angle.

Vertical angles, which are angles across from one another when two lines cross, are always equal. In the illustration below, ∠a = ∠c and ∠b = ∠d.

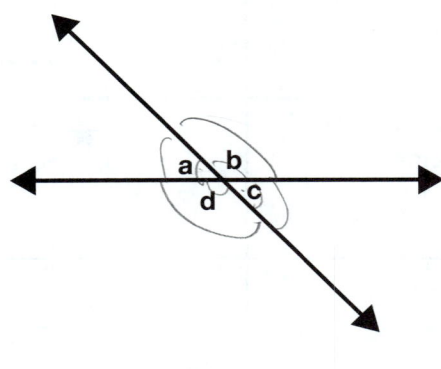

Sometimes parallel lines are intersected by another line. This intersecting line is called a **transversal** and it creates eight angles, four of which are acute and four of which are obtuse (unless the transversal is perpendicular to the lines). Look at these parallel lines intersected by a transversal:

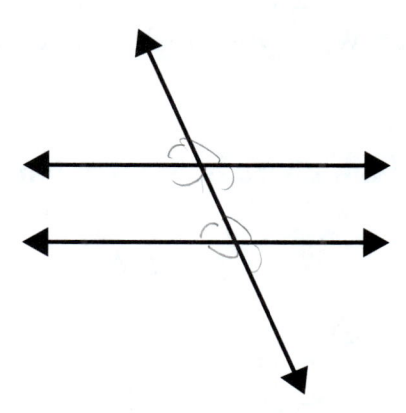

When parallel lines are intersected by a transversal, the four acute angles are always equal and the four obtuse angles are always equal. Each acute and obtuse angle together form supplementary angles, whose sum is 180°.

Let's Review 8: Lines and Angles

Complete each of the following questions. Use the Tip below each question to help you choose the correct answer. When you finish, check your answers with those at the end of Chapter 5.

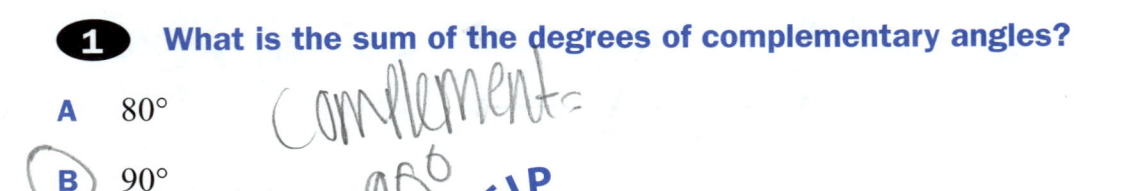

1 What is the sum of the degrees of complementary angles?

A 80°

B 90°

C 120°

D 180°

TIP: If you're not sure of the total measure of complementary angles, reread the information in the previous section of this chapter.

2 Two streets in Josh's neighborhood run next to each other in the same direction but do not intersect. These streets are an example of what kind of lines?

F perpendicular

G adjacent

H supplementary

J parallel

TIP: Try to remember the name for two lines that do not intersect.

Triangles

A **triangle** is a plane figure with three sides. Each of the three corner points on a triangle is called a **vertex**. You read earlier in this chapter that the sum of the angles in a triangle is 180°. Triangles can be named according to their shape, as shown in the table below.

TRIANGLES

Triangle	Description	Example
Equilateral triangle	An equilateral triangle has three equal sides and three equal angles; each angle is 60°.	
Isosceles triangle	An isosceles triangle has at least two equal sides and two equal angles. For example, an isosceles triangle might have angles measuring 80°-50°-50°.	
Scalene triangle	A scalene triangle has no equal sides and no equal angles.	
Right triangle	A right triangle has one right angle. The side opposite the right angle is called the hypotenuse. The other two sides are called legs.	

Triangles have the following properties:

- If two angles in a triangle are congruent to two angles in a second triangle, the third angles in both triangles are also congruent.

- There can be at most only one right or one obtuse angle in a triangle.

- The acute angles in a right triangle are complementary.

- If two angles in a triangle are congruent, the sides opposite those angles are also congruent.

- All equilateral triangles are similar to each other.

Pythagorean Theorem

The **Pythagorean theorem** is a formula used to find the length of one side of a right triangle when you know the lengths of the other two sides. The formula is

$$a^2 + b^2 = c^2$$

where a and b are the lengths of the legs and c is the length of the hypotenuse. The letters a, b, and c are most commonly used, but any variables can be used for the lengths of the legs or hypotenuse.

Look at the triangle below.

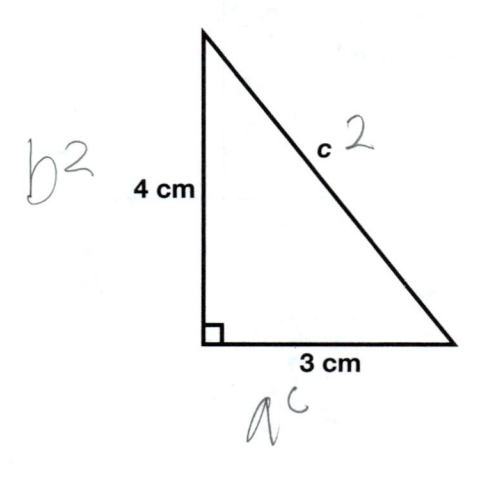

Chapter 5: Geometry and Spatial Sense, Part 1   125

To find the length of the hypotenuse in this triangle, substitute 3 cm and 4 cm as a and b, respectively, into the formula $a^2 + b^2 = c^2$:

$$3^2 + 4^2 = c^2$$
$$9 + 16 = 25$$

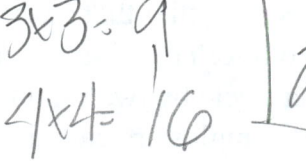

Then find the square root of 25, which is 5.

The length of the hypotenuse is 5 cm.

If you are given the length of the hypotenuse, but are missing the length of either leg, you can still use the Pythagorean Theorem to find the missing side. Look at this triangle:

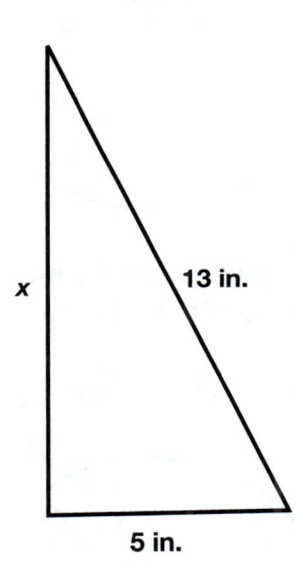

$$5^2 + x^2 = 13^2$$
$$13^2 - 5^2 = x^2$$
$$x^2 = 169 - 25$$
$$x^2 = 144$$
$$x = \sqrt{144} = 12 \text{ in.}$$

Circles

We saw earlier that a circle has a **center**, which is the same distance from any point on the circle. This distance is called the **radius** (plural is *radii*). Angles can be formed inside a circle by two radii that meet at the center, which is the vertex. When this happens, the angles are called **central angles**. The sum of any central angle and its reflex angle in a circle is 360°. The central angle in half a circle is 180°. Look at the circle below. It has two central angles that are labeled A and B. A third central angle, a reflex angle, is unmarked.

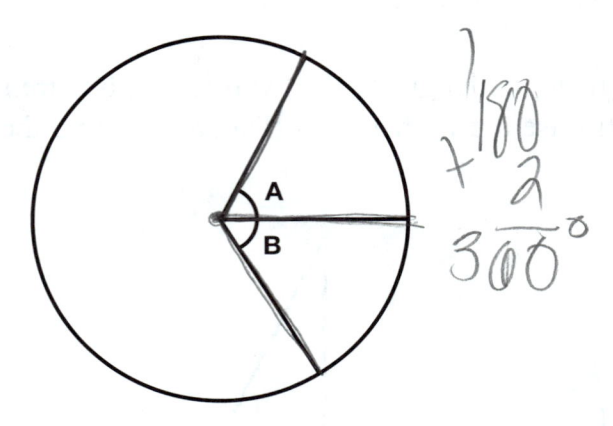

In the following circle, you can figure out the measure of the third central angle by subtracting the sum of the measures of the other two angles from 360°:

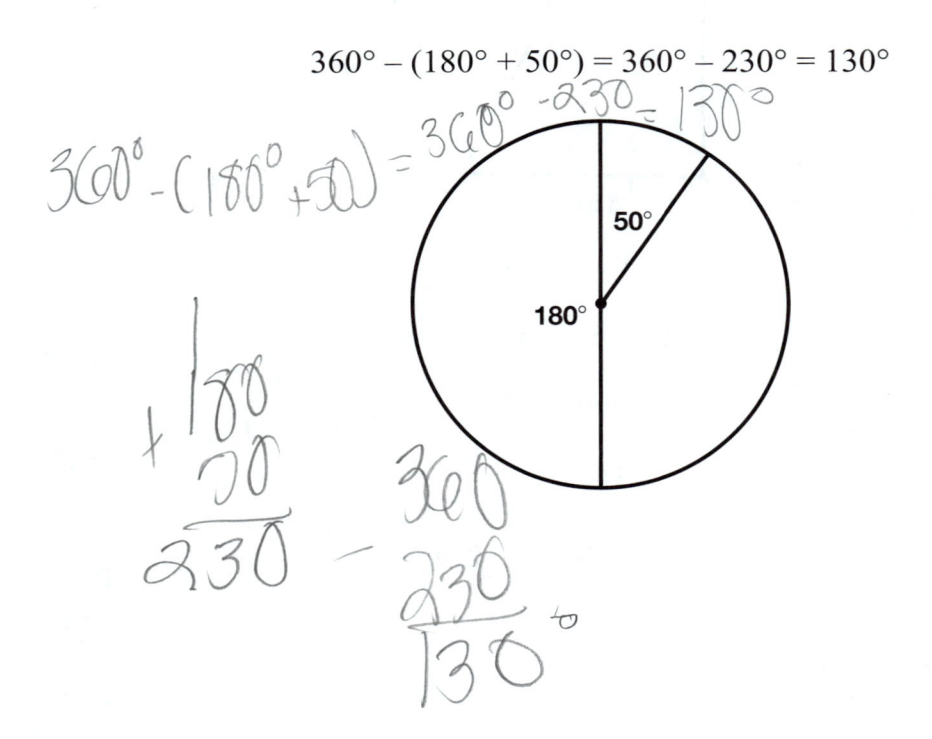

$$360° - (180° + 50°) = 360° - 230° = 130°$$

Chapter 5: Geometry and Spatial Sense, Part 1   127

Let's Review 9: Triangles and Circles

Complete each of the following questions. Use the Tip below each question to help you choose the correct answer. When you finish, check your answers with those at the end of Chapter 5.

1

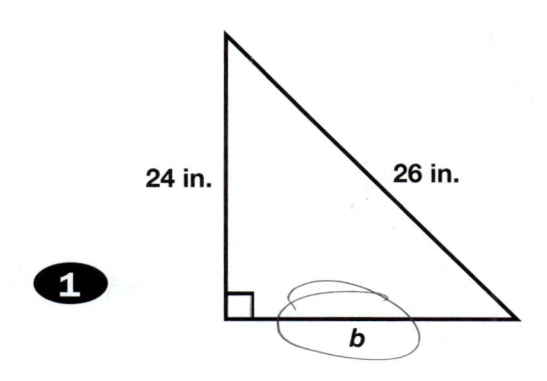

Triangle with legs 24 in. and b, hypotenuse 26 in.

In the right triangle above, the length of one leg equals 24 inches. The length of the hypotenuse is 26 inches. Which formula would determine the length of side *b*?

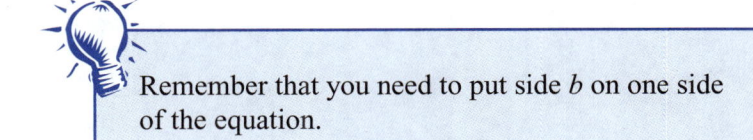

A $24^2 + 26^2 = b^2$

B $b^2 = 24^2 - 26^2$

C $24^2 - b^2 = 26^2$

D $26^2 - 24^2 = b^2$

TIP
Remember that you need to put side *b* on one side of the equation.

2 Beth wants to make a design with a circle divided into pie-shaped pieces of equal size. What is the number of pieces Beth can have if she wants the central angles to be right angles?

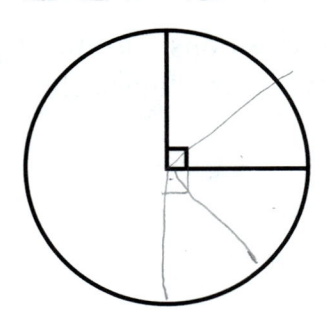

F 2
G 3
H 4
J 5

TIP The pie-shaped piece is one-quarter of the size of the circle.

Chapter 5 Practice Problems

Complete each of the following practice problems. Check your answers at the end of this chapter. Be sure to read the answer explanations!

1 The hexagon below is regular. What is the measure of each of its angles?

A 90°

B 108°

C 120°

D 180°

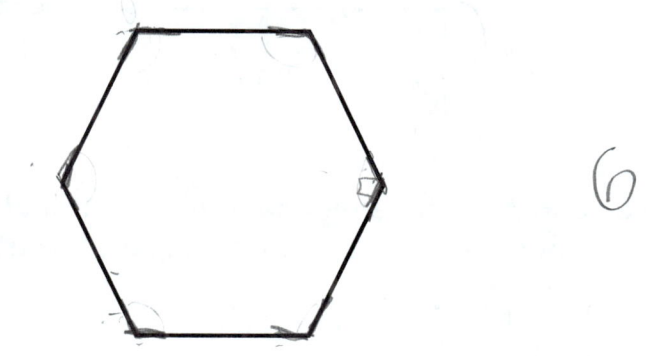

2 An irregular octagon has a perimeter of 64. Seven of its sides measure 4, 4, 9, 8, 8, 5, and 10. What is the length of the remaining side?

F 10

G 12

H 14

J 16

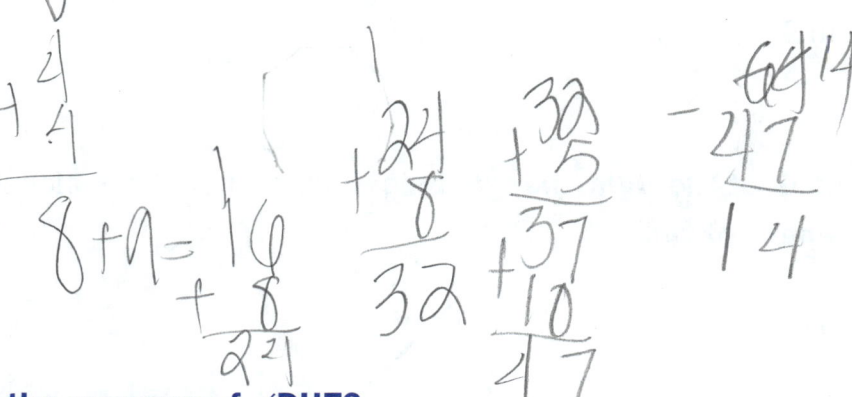

3 What is the measure of ∠DHE?

A 60°

B 90°

C 180°

D 360°

130 TAKS Mathematics

4 Elm Street and Maple Street are parallel to each other. Walnut Street crosses Elm Street and Maple Street as shown in the figure. What is the measure of $\angle a$?

F 20°
G 25°
(H) 75°
J 80°

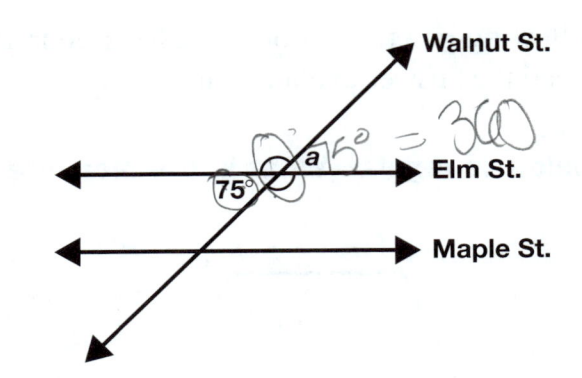

5 Two streets in Terry's neighborhood intersect and form four right angles. These streets are an example of what kind of lines?

(A) perpendicular
B adjacent
C supplementary
D parallel

6 Look at the right triangle below. What number is closest to the length of side c?

F 4 cm
(G) 6 cm
H 8 cm
J 16 cm

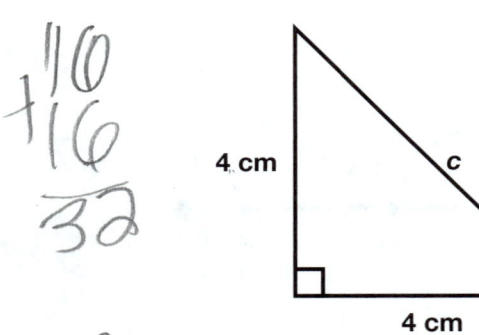

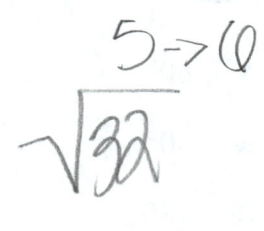

Chapter 5 Answer Explanations

Let's Review 7: Plane Figures

1. B
The ratio between the picture and the poster is 12 : 24 = 1 : 2. If the picture is 24 inches long, the poster is 48 inches = 4 feet long.

2. H
The corresponding side on the smaller figure is 5 and the ratio between the two figures is 1 : 2, so the correct answer is 10.

3. C
The ratio for these triangles is 3 : 9 = 1 : 3, and the corresponding side of the smaller triangle is 8, so the correct answer is 24.

4. J
A rhombus has no right angles, a trapezoid has at most two right angles, an octagon has at most three right angles, and a rectangle has four right angles.

Let's Review 8: Lines and Angles

1. B
The sum of complementary angles is 90 degrees.

2. J
Lines that do not intersect are parallel.

Let's Review 9: Triangles and Circles

1. D

Since you know the measure of a leg and the hypotenuse, you would subtract the squares to find b^2.

2. H

Beth's design is $\frac{1}{4}$ of the circle.

Chapter 5 Practice Problems

1. C

A regular hexagon has angles that measure 120°.

2. J

To find this answer, add all of the sides of the octagon and subtract from the perimeter.

3. C

The measure of a straight angle is 180°.

4. H

Angle *a* is congruent to the angle labeled 75°.

5. A

Two lines that intersect to form four right angles are perpendicular.

6. G

If you use the Pythagorean theorem, you'll see that the length of side *c* is $\sqrt{32}$ cm = 5.6 cm, which is closest to 6 cm.

Chapter 6

Geometry and Spatial Sense, Part 2

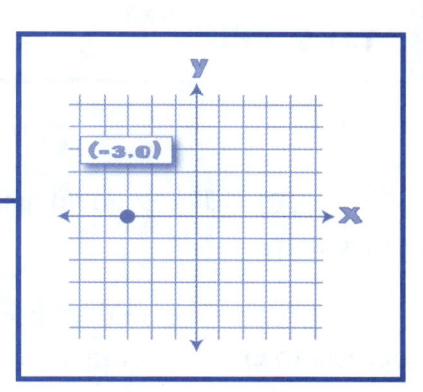

Objectives

Objective 6: The student will demonstrate an understanding of geometric relationships and spatial reasoning.

G(c)(1) **Geometric patterns.** The student identifies, analyzes, and describes patterns that emerge from two- and three-dimensional geometric figures.

(B) The student uses the properties of transformations and their compositions to make connections between mathematics and the real world in applications such as tessellations or fractals.

G(e)(3) **Congruence and the geometry of size.** The student applies the concept of congruence to justify properties of figures and solve problems.

(A) The student uses congruence transformations to make conjectures and justify properties of geometric figures.

Objective 7: The student will demonstrate an understanding of two- and three-dimensional representations of geometric relationships and shapes.

G(d)(1) **Dimensionality and the geometry of location.** The student analyzes the relationship between three-dimensional objects and related two-dimensional representations and uses these representations to solve problems.

(B) The student uses nets to represent (and construct) three-dimensional objects.

(C) The student uses top, front, side, and corner views of three-dimensional objects to create accurate and complete representations and solve problems.

G(d)(2) **Dimensionality and the geometry of location.** The student understands that coordinate systems provide convenient and efficient ways of representing geometric figures and uses them accordingly.

(A) The student uses one- and two-dimensional coordinate systems to represent points, lines, line segments, and figures.

(B) The student uses slopes and equations of lines to investigate geometric relationships, including parallel lines, perpendicular lines, and [special segments of] triangles and other polygons.

(C) The student [develops and] uses formulas, including distance and midpoint.

G(e)(2) **Congruence and the geometry of size.** The student analyzes properties and describes relationships in geometric figures.

(D) The student analyzes the characteristics of three-dimensional figures and their component parts.

In this chapter, you'll learn about three-dimensional figures and coordinate planes. Some of these questions will involve the movement of objects on a coordinate plane. You'll also learn how to calculate the slope of a line and how nets can be used to build three-dimensional figures.

Three-Dimensional Figures

In the previous chapter of this book, you learned about two-dimensional figures called plane figures. In this chapter, you'll learn about three-dimensional figures, which have depth (they are not flat). It is a good idea to learn the properties of the three-dimensional figures summarized in the following table before taking the TAKS. Notice that there are special terms for parts of three-dimensional shapes. Each flat part on a three-dimensional shape is called a **face**. The flat parts are actually two-dimensional shapes as discussed above. The segments where faces meet are called **edges**, and the edges meet at **vertices**, or **corners**. (Vertices is the plural of **vertex**.)

THREE-DIMENSIONAL FIGURES

Shape	Description	Example
Rectangular Solid	Box with six rectangular faces and eight vertices; each corner is a right angle, and the opposite faces are congruent.	
Square Pyramid	Base is a square; has four triangular faces that meet at a vertex.	
Sphere	Ball shape for which every point on the surface is the same distance from the center. It has no vertices or edges.	
Right Circular Cone	Base is circular; top is shaped like a cone with the vertex directly above the center of the circular base; the distance from the vertex to the circular base is the height.	
Right Circular Cylinder	Top and bottom are parallel circles; height is the distance from top to bottom.	
Triangular Prism	Parallel bases are congruent triangles; has three rectangular faces and six vertices.	

The Coordinate Plane

A **coordinate plane** is a graph with four quadrants, I, II, III, and IV. It has an *x*-axis and a *y*-axis. The *x*-axis is horizontal and the *y*-axis is vertical. They intersect at a point called the **origin**. Look at the coordinate plane below. Find the *x*-axis and the *y*-axis and look at the different quadrants.

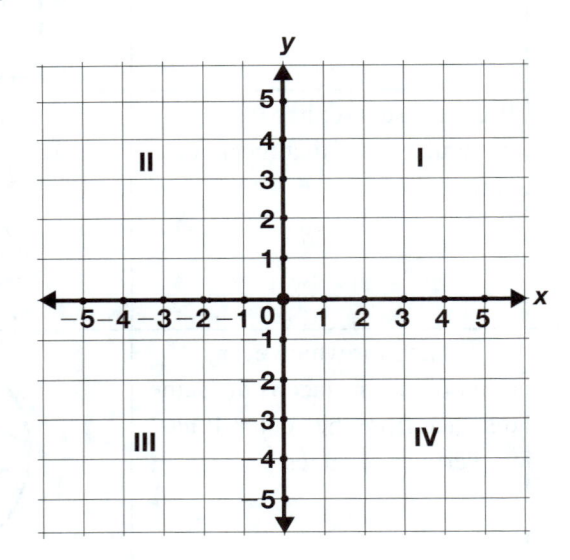

To find the coordinates of a point, move along the *x*-axis first. If the number of the first coordinate is positive, move to the right of the origin. If it's negative, move to the left. Then move along the *y*-axis. If the number is positive, move up from the origin. If it's negative, move down. Look at the coordinate grid shown here. Note that the coordinates of Point A are (4,3). Remember that the *x* value always appears first.

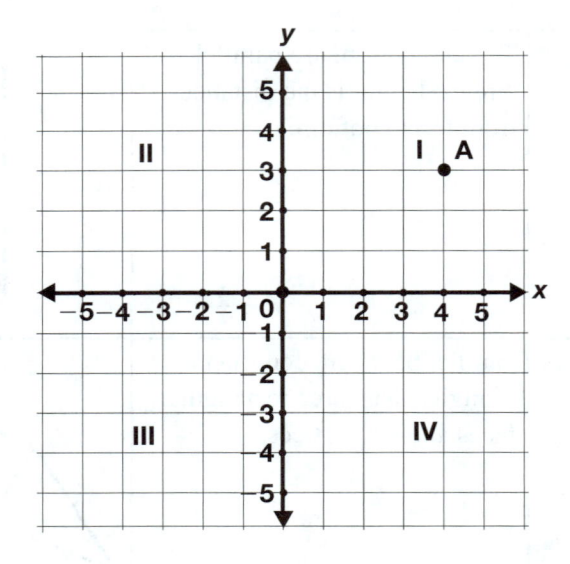

For some test questions, you will have to apply what you've learned about geometric shapes to shapes on a coordinate grid. Read this question:

What is the most accurate name for quadrilateral *ABCD* graphed below?

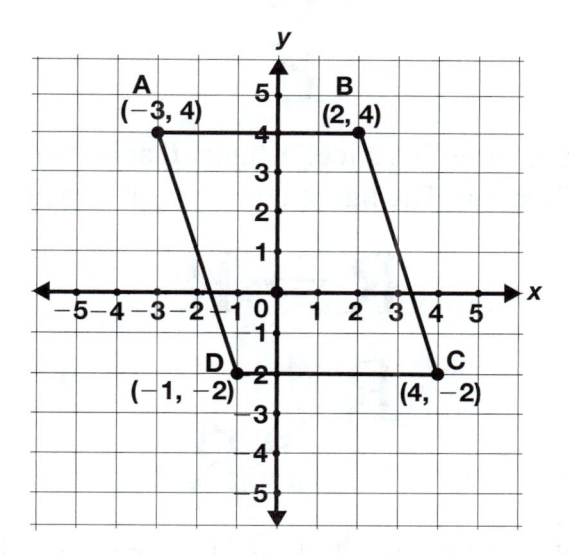

Transformations

Some test questions will be about **transformations**, the movement of figures on a coordinate plane. On the TAKS, you might be asked to choose the correct coordinates of a figure moved in a certain way on the coordinate plane. The following are some common transformations.

1. Rotation: When you rotate a figure, you move it around a fixed point, which is called the **center of rotation**. A rotation can be large or small. Think of the figure being attached to the end of the minute hand of a clock, and you can visualize what a rotated figure looks like. A rotation of 180° is called a half-turn. The figure will look upside down, but not like a mirror image.

A rotation of 90° is called a quarter turn. The figure will look like it is tipped over clockwise on its side.

2. Reflection: When a figure is reflected, it is flipped across a line. The line may be visible or you may have to imagine it. A reflection of a figure is a mirror image.

$$\frac{R}{Я}$$

3. Translation: A translation is a slide. A figure that is translated is moved as if you were sliding it in one direction. The figure does not turn when it is translated.

$$R \rightarrow R$$

$$R \searrow R$$

Other transformation questions on the TAKS may ask you about shapes on a coordinate grid. Read this question:

The following graph shows quadrilateral *TUVW*, and line segments $\overline{AB}$ and $\overline{BC}$.

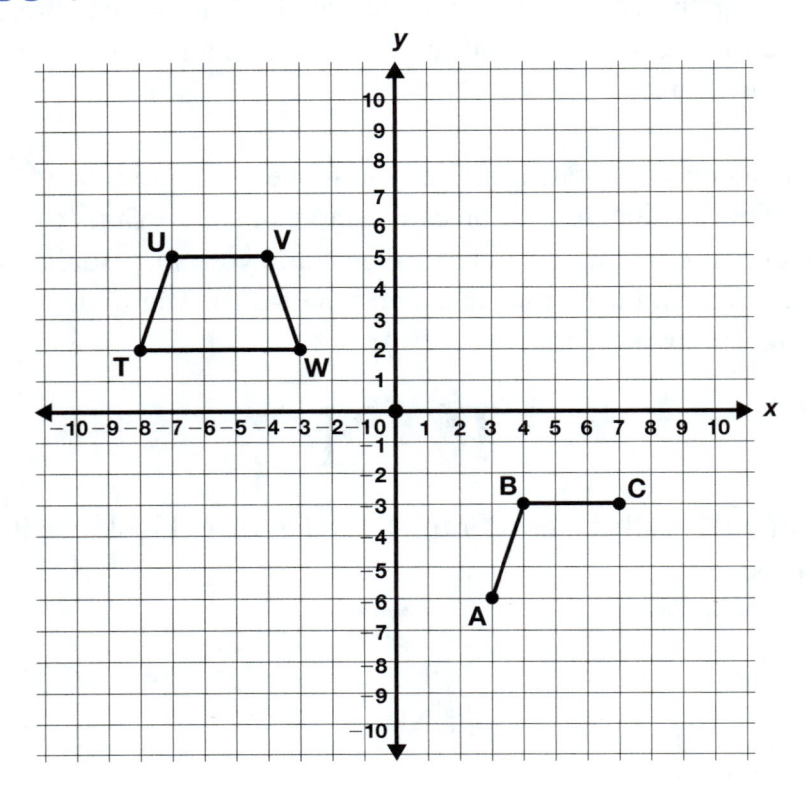

Where should point *D* be placed so that quadrilateral *ABCD* is a translation of *TUVW*?

Remember that congruent figures are exactly the same **size and shape**. Therefore, the correct set of coordinates for point D is (8, −6).

Other questions may be about the movement of shapes not on a coordinate grid. Read this question:

The regular octagon below shows selected positions on a combination lock.

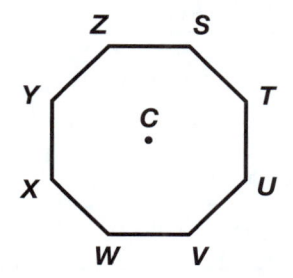

The dial of the lock is turned 45 degrees clockwise and then 135 degrees counterclockwise. Which pair of points can describe the starting and ending points of a marker on the dial of this lock?

A   T to U

B   S to V

C   W to U

D   Z to W

To answer this question, you need to remember that a full rotation is 360 degrees, so each of the eight points on the dial represents one-eighth of 360 degrees, or 45 degrees. So the lock is moved 45 degrees and then back 135 degrees, and W to U is the only answer choice that reflects this rotation.

Let's Review 10: Three-Dimensional Figures

Complete each of the following questions. Use the Tip below each question to help you choose the correct answer. When you finish, check your answers with those at the end of Chapter 6.

1 Which of these sets of ordered pairs form the vertices of an isosceles triangle?

A (0, 2), (3, 5), and (6, 2)

B (−3, 4), (−3, 7), and (1, 7)

C (0, 2), (1, 7), and (3, 5)

D (6, 2), (7, 6), and (8, 5)

TIP: Remember that an isosceles triangle has at least two equal sides.

2 Moving a geometric figure around a fixed point is transformation by

F Inversion

G Reflection

H Rotation

J Translation

TIP: Choose the type that moves the figure around a point.

Chapter 6: Geometry and Spatial Sense, Part 2   141

3 Study Figures I and II. Determine which transformation, if any, of Figure I is shown in Figure II.

$$A \rightarrow A$$

Figure I   Figure II

- **A** Rotation
- **B** Reflection
- **(C)** Translation
- **D** No transformation

TIP: The figure looks as if it has been slid.

4 Which of the following 3-dimensional figures has only 2 faces?

- **F** Cube
- **G** Cylinder
- **H** Triangular Prism
- **J** Sphere

TIP: If you're not sure, go back and reread the description for each three-dimensional figure in the answer choices.

Nets

A **net** is a plane figure that can be folded into a three-dimensional shape. Most times you will be able to tell by the shape of a net what figure it can be folded into. Other times, you will have to count on other clues, such as how many faces and corners it has. Look at this net:

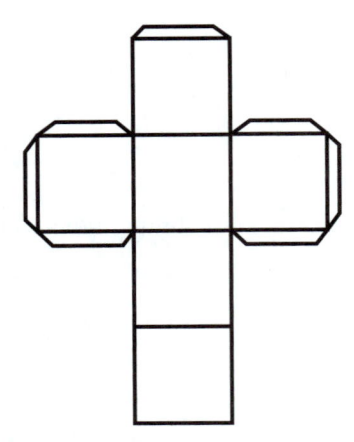

Notice the details in the net. The figure it can be folded into has six faces that are the same size. What three-dimensional figure has six square faces that are the same size? If you said a cube, you're correct!

Alternate Views

You might be asked a question on the TAKS in which you will have to choose a shape that shows how a figure looks from a different angle. Look at the three-dimensional shape below.

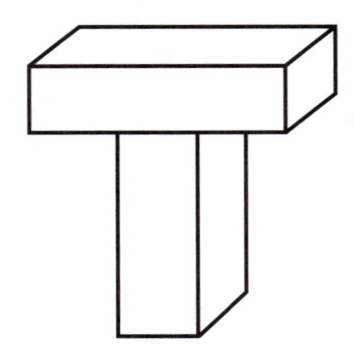

Chapter 6: Geometry and Spatial Sense, Part 2   143

Which picture shows what the shape looks like from the side?

A

B

C

D

To answer this question, you have to think what the figure would look like if you saw it from the side. It would not look like answer choice B, because there would be a dividing line between the horizontal part of the figure and the vertical part. It would not look like answer choice C, which is another front view, or D, which has two dividing lines. It would look like answer choice A.

Chapter 6 Practice Problems

Complete each of the following practice problems. Check your answers at the end of this chapter. Be sure to read the answer explanations!

1 The following graph shows rhombus *ABCD*, and line segments $\overline{KL}$ and $\overline{NK}$.

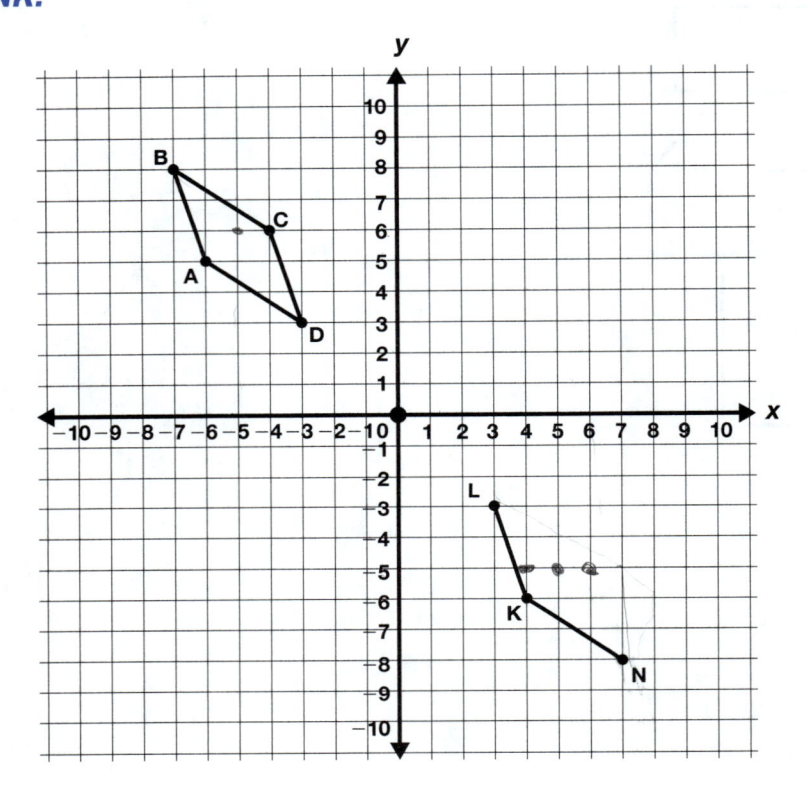

At what coordinate point should vertex *M* be placed to make rhombus *KLMN* congruent to rhombus *ABCD*?

A (−5, 6)

B (6, −5)

C (5, −5)

D (4, −5)

Chapter 6: Geometry and Spatial Sense, Part 2   145

2 Cheryl is designing a wallpaper border. She is reflecting triangle *ABC* over the dashed diagonal line on the coordinate plane to create triangle *A'B'C'*.

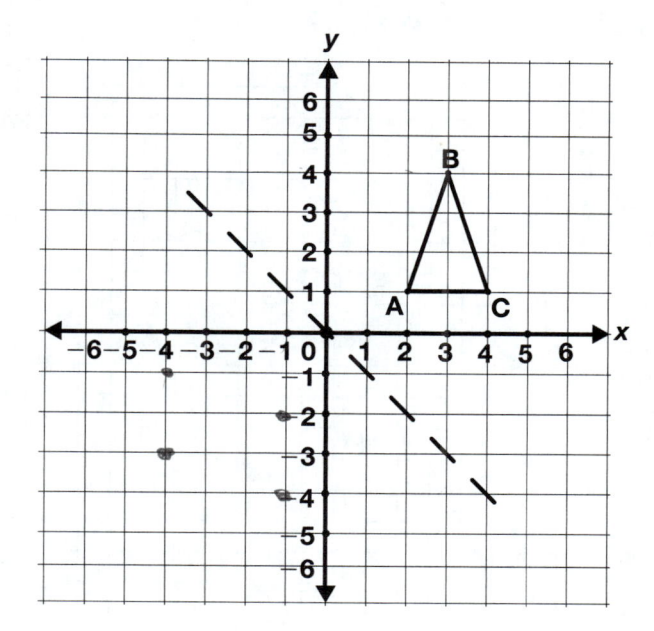

What are the coordinates of *C'*?

F (−1, −2)

G (−1, −4)

H (−4, −3)

J (−4, −1)

3 Which is the slope of a line that passes through the points (2, 4) and (−7, 10)?

A $\dfrac{2}{3}$

B $-\dfrac{2}{3}$

C 2

D −2

4 An art student is making geometric designs for a special project. She plots the coordinates of the vertices of a rectangle on a grid. The first three coordinates are (3,2), (3, 5), and (8, 5). What are the coordinates of the fourth vertex?

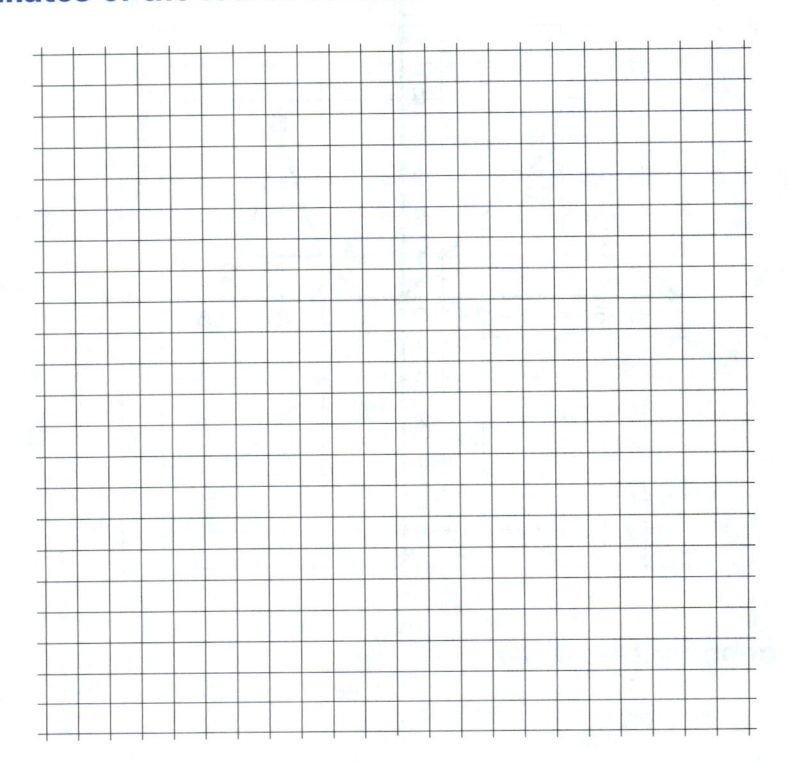

F (2, 8)

G (3, 8)

H (8, 2)

J (8, 3)

Chapter 6: Geometry and Spatial Sense, Part 2 — 147

5 Jane needs to make a model of an ice-cream cone for art class. She plans to cut a shape out of posterboard and then fold it to make a cone with a round base and a pointed top. Which of these nets or shapes could Jane use to make her model?

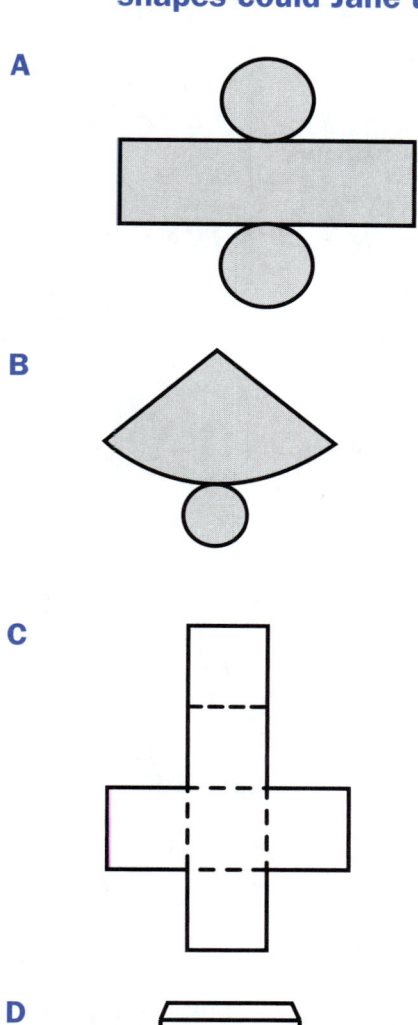

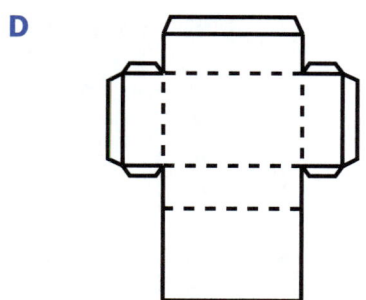

A

B

C

D

6 The illustration below shows a three-dimensional view of a screw.

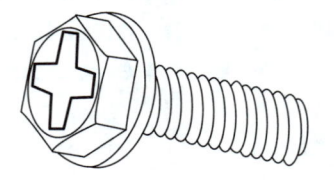

Which of the following is the top view of the screw?

F

G

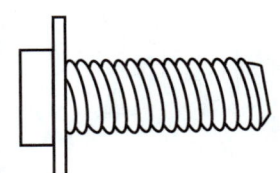

H

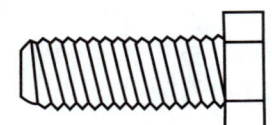

J

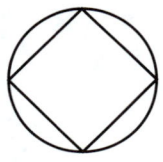

Chapter 6 Answer Explanations

Let's Review 10: Three-Dimensional Figures

1. A

An isosceles triangle has two sides that are exactly the same length. Plot each set of points on the coordinate plane. Only answer choice A gives the coordinates of an isosceles triangle.

2. H

If you weren't sure of this answer, you could figure it out using process of elimination. It is not a reflection or mirror image. Inversion was not discussed, so you can eliminate this answer choice. It's not a translation, which is a slide. It's a rotation.

3. C

When you slide a shape, it's called a translation.

4. G

A cylinder is a three-dimensional figure with only two faces.

Chapter 6 Practice Problems

1. B

For the figure to be congruent to *ABCD,* point *M* should make the slope of side $\overline{LM}$ the same as that of $\overline{BC}$ and $\overline{KN}$. Find this point by moving over three units and down two units from *L*. Likewise, the slope of $\overline{MN}$ should be the same as that of $\overline{CD}$ and $\overline{KL}$. The coordinates thus are (6, −5).

2. G

The coordinates of C' are $(-1, -4)$. In this case, C' can be found by counting down 5 units from c to the line of reflection, then left 5 units from your position on the line.

3. B

If you substitute these coordinates into the formula to find the slope, you'll see that the slope is $-\frac{2}{3}$.

4. H

A rectangle has two pairs of equal and parallel sides. Therefore, the correct coordinates are $(8, 2)$.

5. B

The net shown in answer choice B has a circle for the base, and can be formed into a shape with a point.

6. F

Look carefully at the top of the screw and the direction of the indentations. Answer choice F is the only answer choice that shows the same screw.

Chapter 7

Measurement

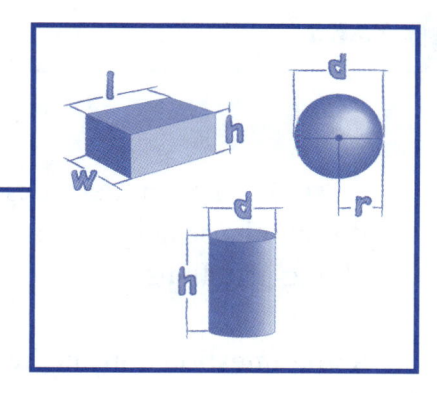

Objectives

Objective 8: The student will demonstrate an understanding of the concepts and uses of measurement and similarity.

G(e)(1) **Congruence and the geometry of size.** The student extends measurement concepts to find area, perimeter, and volume in problem situations.

  (A) The student finds area of regular polygons and composite figures.

  (B) The student finds areas of sectors and arc lengths of circles using proportional reasoning.

  (C) The student [develops, extends, and] uses the Pythagorean theorem.

  (D) The student finds surface area and volumes of prisms, pyramids, spheres, cones, and cylinders in problem situations.

G(f)(1) **Similarity and the geometry of shape.** The student applies the concepts of similarity to justify properties of figures and solve problems.

  (A) The student uses similarity properties and transformations to [explore and] justify conjectures about geometric figures.

  (B) The student uses ratios to solve problems involving similar figures.

  (C) In a variety of ways, the student [develops,] applies, and justifies triangle similarity relationships, such as right triangle ratios, [trigonometric ratios], and Pythagorean triples.

(D) The student describes the effect on perimeter, area, and volume when length, width, or height of a three-dimensional solid is changed and applies this idea in solving problems.

Some questions on the Texas Assessment of Knowledge and Skills (TAKS) will be about measurement. To answer these questions, you will need to apply what you've learned in earlier chapters about plane and three-dimensional figures. You'll learn how to determine the perimeter and area of plane figures and the volume and surface area of three-dimensional figures in this chapter.

Perimeter

The **perimeter** is the distance around a plane figure. Remember that a **plane figure** is a flat, closed figure. The perimeter is commonly measured in inches, feet, centimeters, and meters. The perimeter of a circle is called the **circumference**.

Find the perimeter of the hexagon shown below:

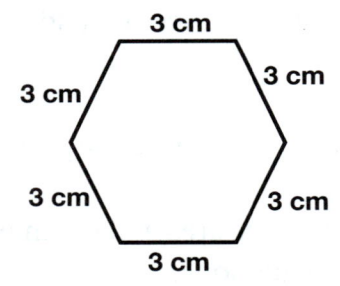

To find the perimeter, add all of the sides $3 + 3 + 3 + 3 + 3 + 3 = 18$, or $6 \times 3 = 18$.

Read this question:

A regular hexagon has a perimeter of 36 inches. What is the length of each of its sides?

To answer this question, you need to know that a hexagon has six sides. If the hexagon is regular, the sides are equal in length. If the perimeter is 36, you can divide it by 6 (the number of sides) to find the length of the sides. The sides are 6 inches long.

Circumference

You just learned that the perimeter of a circle is called the **circumference**. Use this formula to find a circle's circumference: $C = \pi \times$ diameter, where $\pi \approx 3.14$. Look at this circle:

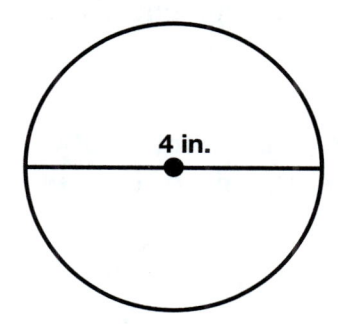

4 in.

To find the circumference, multiply π (3.14) times the diameter (4). Then round to the nearest whole number. The circumference is approximately 13 inches.

Sometimes only the radius of a circle is given. When this happens, you have to double the radius since the diameter is twice the length of the radius. Look at the circle below.

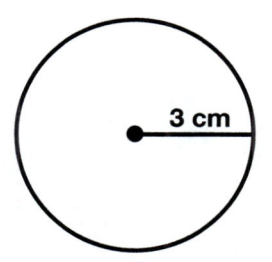

3 cm

To find the circumference of this circle, first double the radius: $3 \times 2 = 6$. Then plug the numbers into the formula $C = \pi \times$ diameter.

$$C = (3.14) \times 6$$
$$C \approx 19 \text{ cm}$$

Let's Review 11: Perimeter

Complete each of the following questions. Use the Tip below each question to help you choose the correct answer. When you finish, check your answers with those at the end of Chapter 7.

1 The perimeters of the two triangles shown below are equal. What is the length of the missing side of the second triangle?

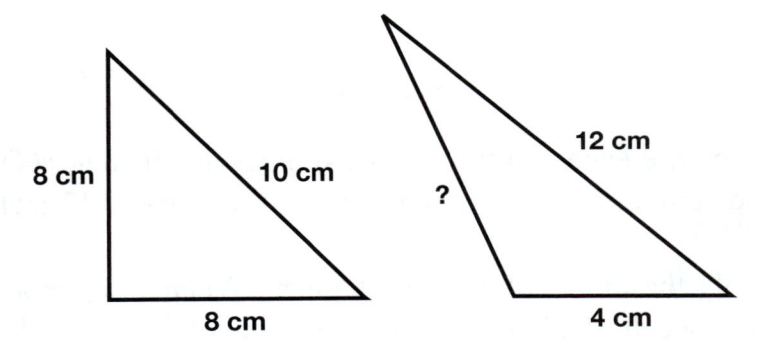

- A 4 cm
- B 6 cm
- C 8 cm
- D 10 cm

TIP

Remember that the perimeter of a triangle is the sum of all three sides.

Chapter 7: Measurement   155

2 A hexagon has a perimeter of 25. Five of its sides are 3, 3, 4, 6, and 6. What is the length of the remaining side?

F   2

G   3

H   4

J   5

TIP
The perimeter is the sum of all six sides.

3 The regular hexagon below has the same perimeter as a square with a side of 18 inches. How long is each side of the hexagon?

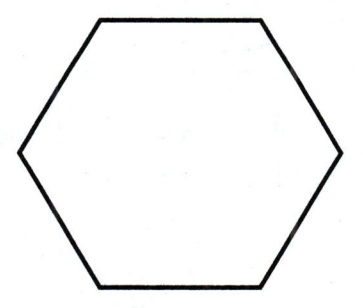

A   10 inches

B   12 inches

C   24 inches

D   72 inches

TIP
All the sides of a regular hexagon are equal, and all the sides of a square are equal.

4 Kelly has an irregularly shaped backyard as shown below.

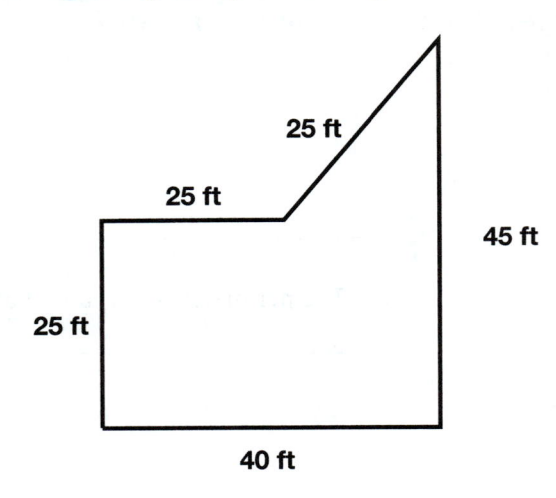

What is the perimeter of his backyard?

F 150 feet

G 160 feet

H 170 feet

J 180 feet

TIP: Add together the sides of Kelly's yard to solve this problem.

Area

For some questions on the TAKS you will have to determine the area of a figure. These formulas listed here for area are also on your reference sheet:

Rectangle   Area = length × width or Area = base × height;
$A = lw$ or $A = bh$

Triangle   Area = $\frac{1}{2}$ base × height or Area = base × height ÷ 2;
$A = \frac{1}{2}bh$ or $A = \frac{bh}{2}$

Trapezoid   Area = $\frac{1}{2}$ height × (base 1 + base 2) or
Area = (base 1 + base 2) × (height ÷ 2);
$A = \frac{1}{2}h(b_1 + b_2)$ or $A = (b_1 + b_2)\left(\frac{h}{2}\right)$

Parallelogram   Area = base × height, or $A = bh$

Circle   Area = pi × radius squared, or $A = \pi r^2$

Look at the rectangle below:

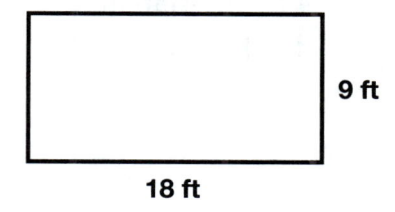

[Rectangle: 18 ft by 9 ft]

To find the area of this rectangle, put its length and width into the formula you just learned:

$$A = lw$$
$$A = 18 \text{ ft} \times 9 \text{ ft}$$
$$A = 162 \text{ ft}^2$$

Notice that the area is expressed in square feet.

To find the area of a circle, use this formula:

$$A = \pi r^2$$

Look at this circle:

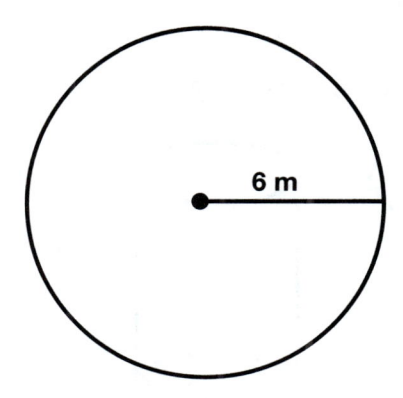

[Circle with radius 6 m]

To find the area of this circle, put the radius into the formula you just learned, which is usually given to you on the test. Use 3.14 for π and round your answer to the nearest whole number.

$$A = (3.14)(6 \text{ m})^2$$
$$A = (3.14)36 \text{ m}^2$$
$$A \approx 113 \text{ m}^2$$

Volume

Three-dimensional figures are different from plane figures because they have width. Unlike plane figures, three-dimensional figures are not flat. It is a good idea to learn the properties of the following three-dimensional figures:

Rectangular prism

Box with six rectangular faces; each corner is a right angle.

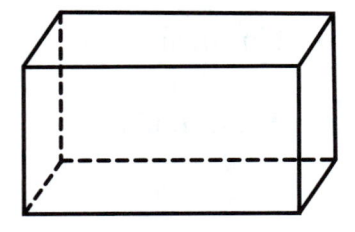

Right Circular Cylinder

Top and bottom are parallel circles; height is the distance from top to bottom.

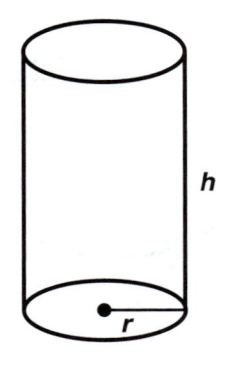

Square Pyramid

Has a base that is a square; has four triangular faces that meet at a vertex.

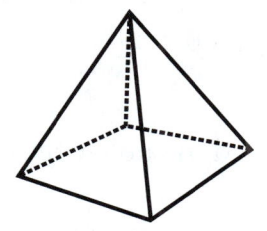

Sphere

Ball where every point on the surface is the same distance from the center.

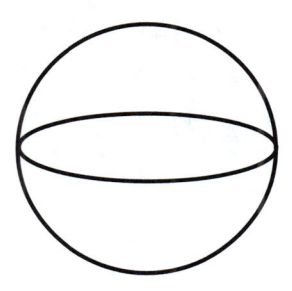

To find the volume, or capacity, of a three-dimensional figure, such as a rectangular prism or right circular cylinder, use these formulas, which are also on your reference sheet:

Prism or Cylinder v = base area × height or $v = Bh$
Rectangular prism V = length × width × height, or $V = lwh$
Right circular cylinder V = (pi) × radius² × height, or $V = \pi r^2 h$

20 cm
30 cm
20 cm

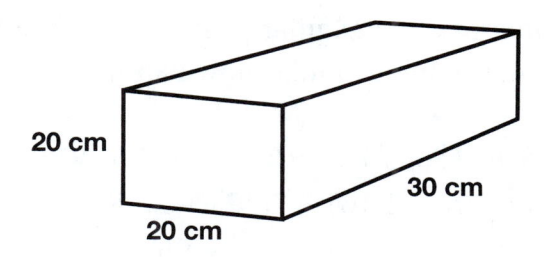

To find the volume of the rectangular prism above, substitute the measurements of the length, width, and height into the formula you just learned:

$$V = 30 \text{ cm} \times 20 \text{ cm} \times 20 \text{ cm}$$
$$V = 12{,}000 \text{ cm}^3$$

Note that the volume is expressed in cubic centimeters.

To find the volume of a cylinder, you would use the formula $V = \pi r^2 h$. To find the volume of a cylinder with a radius of 3 and a height of 5, substitute these values into the formula and round your answer to the nearest whole number.

$$V = (3.14) \times 3^2 \times 5$$
$$V = (3.14) \times 9 \times 5$$
$$V \approx 141$$

Surface Area

To find the total surface area for a rectangular prism or a right circular cylinder, use these formulas, which are also on your reference sheet:

Rectangular prism $S = 2(lw) + 2(hw) + 2(lh)$

Right circular cylinder $S = 2\pi rh + 2\pi r^2$ or $S = 2\pi r(h + r)$

To find the surface area of a rectangular prism with a height of 6 cm, a length of 8 cm, and a width of 2 cm, use the formula shown above:

$$S = 2(8 \times 2) + 2(6 \times 2) + 2(8 \times 6)$$
$$S = 2(16) + 2(12) + 2(48)$$
$$S = 32 + 24 + 96$$
$$S = 152 \text{ cm}^2$$

Chapter 7: Measurement

Let's Review 12: Area

Complete each of the following questions. Use the Tip below each question to help you choose the correct answer. When you finish, check your answers with those at the end of Chapter 7.

1 What is the volume of the box pictured below?

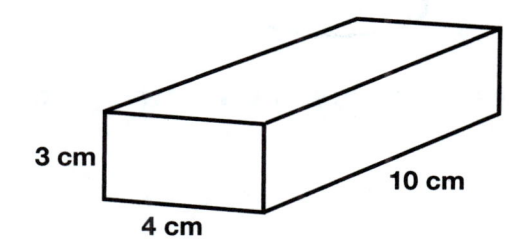

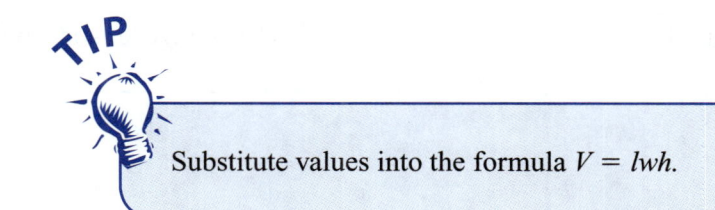

3 cm
4 cm
10 cm

- A 17 cm³
- B 40 cm³
- C 70 cm³
- D 120 cm³

TIP: Substitute values into the formula $V = lwh$.

2 A rotating sprinkler is used to water a yard. The radius of the area being sprayed is 8 feet. What is the approximate wet area of the yard?

- F 20 ft²
- G 25 ft²
- H 64 ft²
- J 201 ft²

TIP: Remember to square the radius.

3 The storage tank below has a radius of 8 feet and a height of 10 feet.

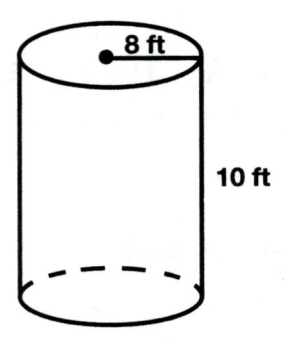

What is the total surface area of the storage tank?

A 402 ft²

B 455 ft²

C 502 ft²

D 904 ft²

TIP The storage tank is a right circular cylinder.

Chapter 7: Measurement 163

Chapter 7 Practice Problems

Complete each of the following practice problems. Check your answers at the end of this chapter. Be sure to read the answer explanations!

1 What is the volume of the box pictured below?

A 15 cm³

B 25 cm³

C 125 cm³

D 250 cm³

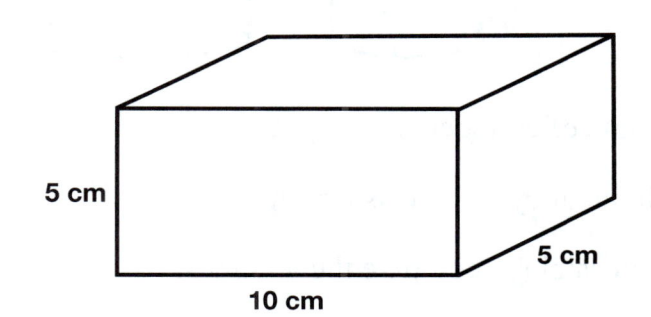

2 Emile wants to wrap the box below with wrapping paper. How many square inches of wrapping paper does he need?

F 100 in.²

G 160 in.²

H 340 in.²

J 420 in.²

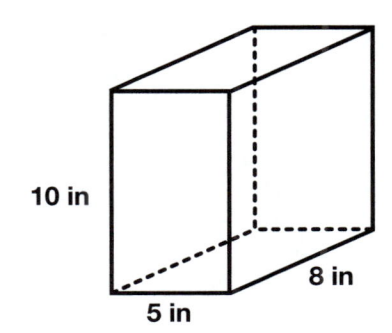

3 The volume of a cylinder is found by using the formula $V = \pi r^2 h$. How do the volumes of cylinder A and cylinder B in the following figure compare?

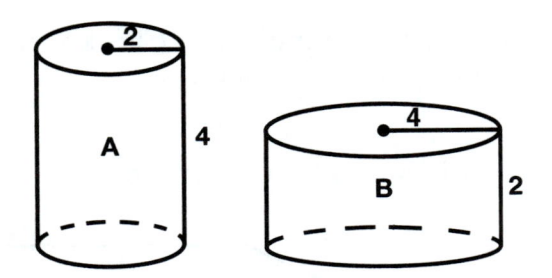

A The volume of cylinder A is larger.

B The volume of cylinder B is larger.

C It is not possible to compare the volumes.

D The volumes of cylinder A and cylinder B are the same.

4 What is the area of a circle with a radius of 10 cm?

F 3.14 cm²

G 31.4 cm²

H 314 cm²

J 3140 cm²

Chapter 7: Measurement 165

5 What is the approximate volume of the box pictured below?

A 17 cm³

B 13 cm³

C 36 cm³

D 127 cm³

3.5 cm

3.9 cm

9.3 cm

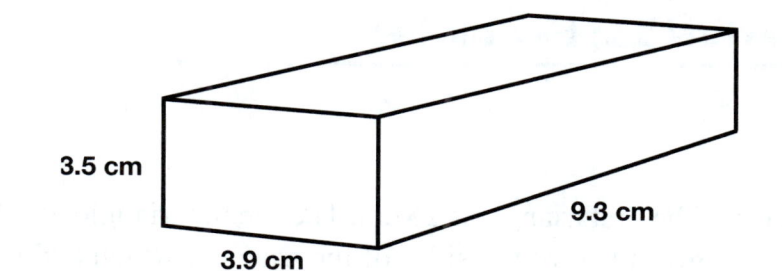

6 Cheyenne made a triangular prism in art class. The net of the prism is shown below. Each of the three faces is a square with a side of 2 cm. and each of the bases is an isosceles right triangle.

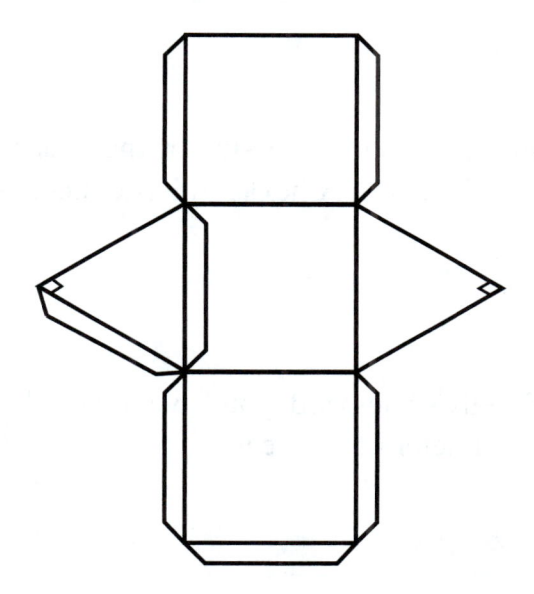

Which is closest to the volume of this triangular prism (use $V = Bh$)?

F $\frac{1}{2}$ cm³

G 1 cm³

H 2 cm³

J 4 cm³

Chapter 7 Answer Explanations

Let's Review 11: Perimeter

1. D
The perimeter of the first triangle is 26 cm. The second triangle has the same perimeter, but we are only given two of the sides of the triangle, which add up to 16. Therefore, the last side must be 10 cm.

2. G
The sides of the hexagon add up to 22 and its perimeter is 25. Therefore, the missing side must be 3.

3. B
All sides on a square are equal, so if one side of the square is 18 inches long, the perimeter of the square is 72 inches. A hexagon has 6 sides, so you need to divide 6 into 72, which is 12.

4. G
To find the perimeter of Kelly's backyard, you have to add the sides: 25 + 25 + 25 + 45 + 40. His yard has a perimeter of 160 feet.

Let's Review 12: Area

1. D
When you multiply 10 by 4 by 3, you get 120.

2. J
To solve this problem, you need to find the area of a circle with a radius of 8 feet. If you substitute $r = 8$ ft into the formula $A = \pi r^2$, the answer is 201 ft^2.

3. D

Substitute values into the formula $S = 2\pi rh + 2\pi r^2$ to find the surface area of a right cylinder:

$$2 \times (3.14) \times 8 \times 10 + 2 \times (3.14) \times 8^2$$
$$6.28 \times 80 + 6.28 \times 64$$
$$\approx 502 + 402 = 904 \text{ ft}^2.$$

Chapter 7 Practice Problems

1. D

Substitute the values in the formula for the volume of rectangular solid: $V = lwh$:

$$V = 5 \times 5 \times 10$$
$$V = 250 \text{ cm}^3.$$

2. H

To determine how much wrapping paper Emile needs, substitute the dimensions of the box into the formula to find the surface area of a rectangular prism:

$$S = 2(lw) + 2(hw) + 2(lh)$$
$$S = 2(8 \times 5) + 2(10 \times 5) + 2(8 \times 10)$$
$$S = 2(40) + 2(50) + 2(80)$$
$$S = 80 + 100 + 160$$
$$S = 340 \text{ in.}^2$$

3. B

The volume of cylinder A is about 50; the volume is cylinder B is about 100.

4. H

When you substitute the radius of 10 cm into the formula $A = \pi r^2$, you get 3.14×100, or about 314 cm².

5. D

When you multiply the length, width, and height of the box, the answer is about 127 cm³.

6. H

Let x represent each leg of an isosceles right triangle. By the Pythagorean theorem, $x^2 + x^2 = 2^2$. Then $2x^2 = 4$, $x^2 = 2$, so $x = \sqrt{2}$. Now the area of the triangular base is $\left(\dfrac{1}{2}\right)(\sqrt{2})(\sqrt{2}) = 1$. The height of the prism is the side of one square, which is 2. Thus, the volume of this prism is $(1)(2) = 2$ cm³.

Chapter 8
Data Analysis and Probability

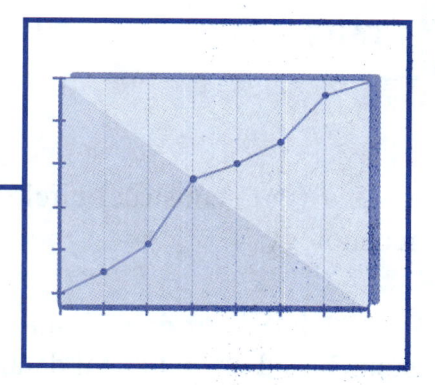

Objectives

Objective 9: The student will demonstrate an understanding of percents, proportional relationships, probability, and statistics in application problems.

(8.3) **Patterns, relationships, and algebraic thinking.** The student identifies proportional relationships in problem situations and solves problems. The student is expected to

(B) estimate and find solutions to application problems involving percents and proportional relationships, such as similarity and rates.

(8.11) **Probability and statistics.** The student applies the concepts of theoretical and experimental probability to make predictions. The student is expected to

(A) find the probabilities of compound events (dependent and independent); and

(B) use theoretical probabilities and experimental results to make predictions and decisions.

(8.12) **Probability and statistics.** The student uses statistical procedures to describe data. The student is expected to

(A) select the appropriate measure of central tendency to describe a set of data for a particular purpose; and

(C) construct circle graphs, bar graphs, and histograms, with and without technology.

(8.13) **Probability and statistics.** The student evaluates predictions and conclusions based on statistical data. The student is expected to

(B) recognize misuses of graphical or numerical information and evaluate predictions and conclusions based on data analysis.

In this chapter, you'll learn how to solve data analysis problems involving probability and measures of central tendency. **Probability** refers to the chance of an event happening. Like many questions on the TAKS, probability problems will often involve real-life situations.

The **measures of central tendency** tested on the TAKS include mean, mode, and median. You'll learn how to determine each of these as well as the range for a set of a data. You'll also learn how to make predictions based on probabilities and interpret information in graphs.

For some questions on the TAKS, you will have to solve real-life problems, such as determining a discount on a sale price. You'll learn how to do this in this chapter.

Probability

The probability of an outcome can be determined by using this formula:

$$P = \frac{\text{number of favorable outcomes}}{\text{number of possible outcomes}}$$

Probability can be expressed as a fraction, a decimal, a percent, or a ratio.

For example, read this problem:

Find the probability of spinning a "D" on the spinner below.

A 0

B $\dfrac{1}{4}$

C $\dfrac{1}{2}$

D 1

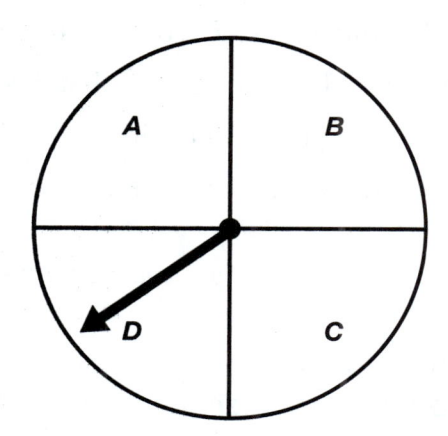

To solve this problem, use the formula shown above, where

1 = the number of favorable outcomes
4 = the number of possible outcomes

Then you can see that the probability of spinning a "D" on the spinner is $\dfrac{1}{4}$. Answer choice B is correct.

Let's try another problem:

Justine has a bag of 20 marbles. Ten of these marbles are white, 3 are green, 2 are blue, and 5 are yellow. If Justine reaches into the bag and pulls out a marble without looking, what is the probability that she will pull out a yellow marble?

F 0

G 5%

H 25%

J 50%

Use the probability formula to solve this problem. There are 5 yellow marbles, so this is the number of favorable outcomes. There are 20 marbles altogether, so this is the number of possible outcomes. The probability that Justine will pull out a yellow marble is $\frac{5}{20}$. If you convert this fraction to a decimal, it's .25. Then if you convert the decimal to a percent, it's 25%. Answer choice H is correct.

Now, let's try a problem that gives you the probability and asks about the outcomes.

There are 10 straws in a box; some are white and some are red. The probability of reaching into the box and selecting a white straw is $\frac{2}{5}$. How many red straws are in the box?

A 1

B 6

C 8

D 10

You have to work backwards to solve this problem. You know there are 10 straws in the box all together, so the denominator must be 10 before it is reduced. You also know that the probability of reaching into the box and pulling out a white straw is $\frac{2}{5}$. To find out how many white straws there are, set up the fractions in the proportion as shown here:

$$\frac{2}{5} = \frac{x}{10}$$

If you multiply $\frac{2}{5}$ by $\frac{2}{2}$, you get $\frac{4}{10}$, which tells you the number of white straws is 4. If there are 10 straws altogether and 4 of them are white, 6 of them must be red. Answer choice B is correct.

Let's Review 13: Probability

Complete each of the following questions. Use the Tip below each question to help you choose the correct answer. When you finish, check your answers with those at the end of Chapter 8.

1 If all central angles are equal, find the probability of spinning "green" on the spinner below.

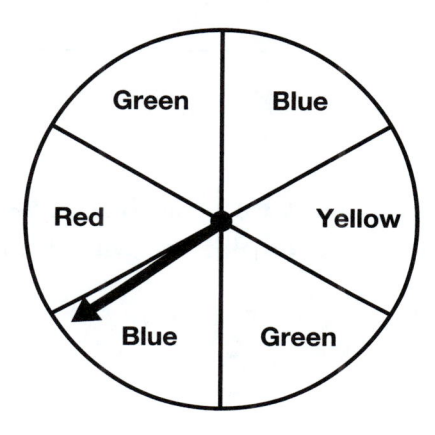

A   0

B $\frac{1}{4}$

C $\frac{1}{3}$

D $\frac{1}{2}$

TIP

Remember to use the formula for probability and then reduce the fraction. There are six sections on the spinner and two of these sections are green.

2 A bag contains 8 white chips, 5 red chips, 3 black chips, 2 blue chips, and 2 green chips. If you reach into the bag without looking what is the probability that you will pull out a red chip?

- F 25%
- G 33%
- H 55%
- J 67%

TIP: Count the total number of chips. Five of these chips are red.

3 Peter is going to roll a six-sided number cube with faces numbered 1 through 6. What is the probability of rolling an even number?

- A $\frac{1}{6}$
- B $\frac{1}{4}$
- C $\frac{1}{3}$
- D $\frac{1}{2}$

TIP: A six-sided number cube has sides numbered 1, 2, 3, 4, 5, and 6.

Chapter 8: Data Analysis and Probability

4 A jar contains only pink and yellow jelly beans. The probability of randomly reaching into the jar and selecting a pink jelly bean is $\frac{1}{4}$. What percentage of beans in the jar are yellow?

F 25%

G 50%

H 60%

J 75%

TIP: Convert $\frac{1}{4}$ to a percentage for the pink jelly beans. Then compute the remaining percentage for the yellow jelly beans.

5 If a penny is tossed 10 times, and on the first five tosses it comes up heads, what is the probability of getting heads on the sixth toss?

A $\frac{1}{4}$

B $\frac{1}{3}$

C $\frac{1}{2}$

D 1

TIP: Think about whether it matters how many times you toss a coin.

Central Tendency: Mean, Median, and Mode

Some questions on the TAKS will ask you to analyze data to find measures of central tendency, including the mean, median, and mode, and to choose the best measure of central tendency to use in a particular situation. **Mean** is another word for average. To find the mean of a set of numbers, add all of the numbers together and divide by the number of items that make up that total. Look at this set of numbers:

$$2, 4, 6, 8, 10$$

To find the mean, you first add all of the numbers:

$$2 + 4 + 6 + 8 + 10 = 30$$

Then you divide 30 by the number of items, in this case 5. The mean of these numbers is therefore 6.

The **median** of a set of numbers is the middle number when the numbers are ordered by size. It's not the average, but simply the number in the middle. Look at this set of numbers:

$$16, 10, 2, 14, 8$$

To find the median, you need to put them in order from least to greatest:

$$2, 8, 10, 14, 16$$

When the numbers are in order from least to greatest, you can see that the number 10 is the median, the middle number.

If there are an even number of values, order them from least to greatest, and the median would be the average (mean) of the two middle numbers. Given the numbers 7, 5, 16, 20, 3, 9, to find the median, first arrange them in order to appear as 3, 5, 7, 9, 16, and 20. The median is the average of the two middle numbers, which are 7 and 9. Thus, the median equals $\frac{(7+9)}{2} = 8$.

Chapter 8: Data Analysis and Probability   177

The **mode** of a set of data is the most frequently occurring number. Look at the numbers below:

$$88, 90, 76, 42, 88, 92, 100, 110, 115$$

The mode of these numbers is 88, the only number that occurs more than once. It is the most frequently occurring number.

Look at the following problem:

If the mean number of people who attended an amusement park over five days was 25,000, what was the total attendance during the five days?

A   5,000

B   50,000

C   125,000

D   250,000

Since the mean is defined as the total divided by the number of items, the total must equal the mean times the number of items. So, to solve this problem, you need to multiply the number of items (days) by the mean. In this case, you would multiply 5 × 25,000. The answer is 125,000.

Range

The **range** of a set of data is the difference between the smallest number and the greatest number. It indicates how spread out the values are. For example, consider these numbers again:

$$88, 90, 76, 42, 88, 90, 100, 110, 115$$

The smallest number is 42 and the greatest is 115. To find the range, subtract 42 from 115:

$$115 - 42 = 73$$

The range of this set of numbers is 73.

As another example, read this problem:

The scores on Mr. Seymour's English test were 98, 60, 88, 87, 96, 79, 80, 58, 76, 99, 80, 58, 76, 99, 90, 87, 62, 76, 89, and 97. What is the range of the scores?

To determine the range of this problem, subtract the lowest test score, 58 from the highest 99.

$$99 - 58 = 41$$

Let's Review 14: Central Tendency

Complete each of the following questions. Use the Tip below each question to help you choose the correct answer. When you finish, check your answers with those at the end of Chapter 8.

1 The office manager in a small office is considering hiring a receptionist to answer the telephone. To see whether a receptionist really is needed, the employees used a log to record the number of calls answered each day. The calls answered each day in a 14-day period are shown below.

10, 12, 8, 16, 6, 5, 8, 5, 12, 13, 12, 12, 8, 6

Which results in the greatest number of calls during the 14-day period?

A Mean

B Mode

C Median

D Range

TIP: Calculate the measure in each answer choice. Then choose the highest number.

2 The number of cars sold at Ray's Used Automobiles was 12 in January, 22 in February, 30 in March, 42 in April, and 58 in May. What is the range in the number of cars sold from January to May?

F 12

G 33

H 46

J 70

TIP: Remember that the range is the difference between the least and the most.

3 The total points scored for the Warriors basketball team for each game during the season were 42, 20, 13, 64, 27, 35, 45, 40, 23, 12, 12, and 39. What is their mean score?

A 12

B 31

C 35

D 52

TIP You need to find the mean of this data. Add the numbers, and then divide by the total number of items.

4 The manager of a retail store recorded the store's sales every day for one week. Which measure did she use to determine that the sales varied by $250 during the week?

F Mean

G Median

H Mode

J Range

TIP In order for the manager to find out how much the sales varied, she would have to subtract the highest sales from the lowest sales.

5 Renee's world cultures grades were 84, 85, 95, 88, 92, 100, 82, and 78. What is her mean grade?

TIP Add together all of her grades, and then divide by 8.

Charts and Graphs

Some questions on the TAKS will be about data displayed in line, bar, and circle graphs. You need to be able to interpret data in these graphs to answer these questions You'll learn about these graphs and plots in this section.

Line Graphs

A **line graph** is a very popular type of graph that compares two variables—one along the *x*-axis and one along the *y*-axis. The two variables being compared in a line graph are closely related, so that a change in one variable causes a change in the second variable. A line graph is a great way to show trends. Look at this line graph:

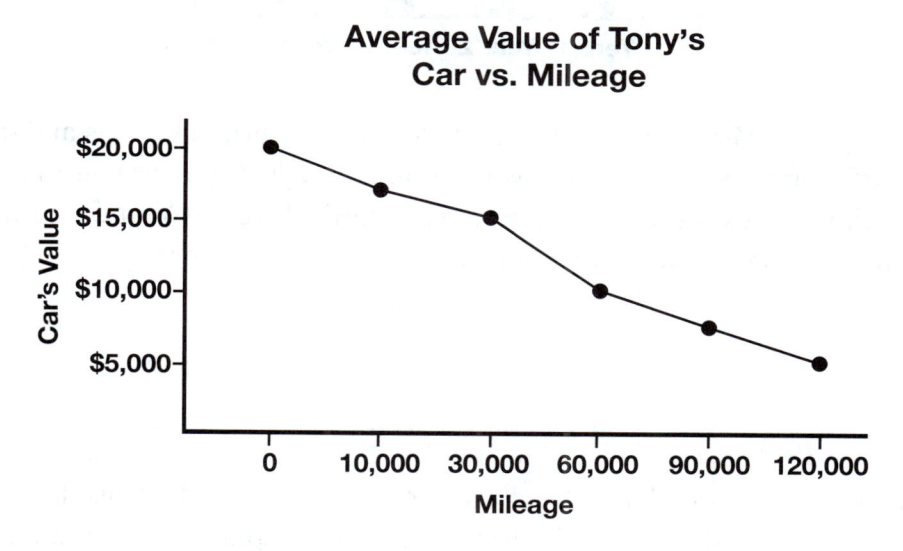

Average Value of Tony's Car vs. Mileage

You can see from this line graph that as the mileage on Tony's car increases, the value of the car decreases.

Bar Graphs

In a **bar graph**, the height or length of a bar indicates a value. The higher or longer the bar, the greater the value. A bar graph has two axes; one indicates what is being measured, and the other indicates the value of that measurement as a bar. It is a good way to show comparisons as well as trends, such as changes in sales over time. It can also display data about unrelated quantities.

A bar graph can have either vertical or horizontal bars, but most bar graphs on the TAKS have vertical bars. Look at this bar graph. It shows the number of cars manufactured at a factory over a five-year period.

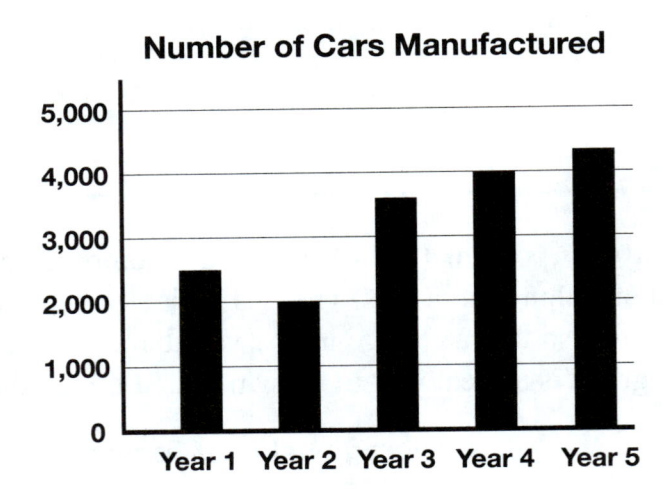

Number of Cars Manufactured

Year 1	~2,500
Year 2	~2,000
Year 3	~3,700
Year 4	~4,100
Year 5	~4,400

In this bar graph, Years 1 through 5 are listed on the horizontal axis and the number of cars manufactured is listed on the vertical axis. By just glancing at the graph, you can see that the greatest number of cars were manufactured in Year 5 and that, except for Year 2, the number of cars manufactured increased each year.

Circle Graphs

A **circle graph**, also called a **pie chart** or **pie graph**, is often used to display the division of a whole. Data in a circle graph are often displayed in percentages. Circle graphs work best to show large divisions, such as the division of money in a household budget. For example, look at the circle graph on the next page:

Chapter 8: Data Analysis and Probability 183

Miller Family Monthly Budget

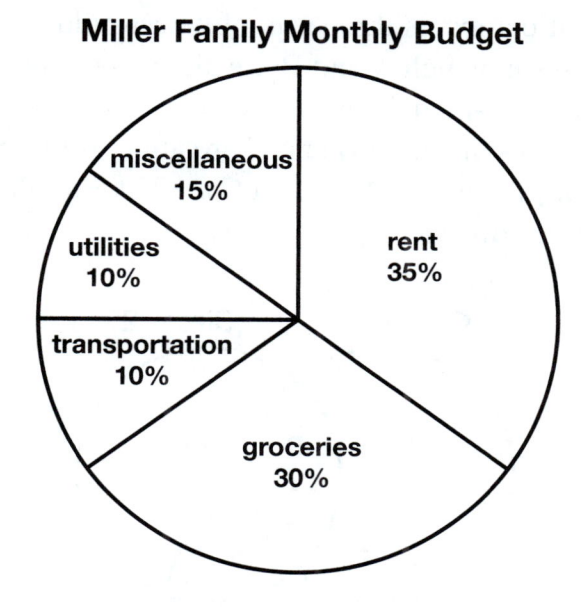

- miscellaneous 15%
- utilities 10%
- rent 35%
- transportation 10%
- groceries 30%

You can see from this circle graph that the Miller family spends most of its monthly income on rent and groceries.

Venn Diagrams

Some questions on the TAKS may ask you to interpret data displayed in a Venn diagram. A **Venn diagram** is made up of two or more overlapping circles and is used to display relationships among two or more sets of data. Venn diagrams are a great way to show similarities and differences. Look at the Venn diagram below. If this diagram were filled in, it would compare two sets of data, A and B. The part of the circle that overlaps, C, would list ways that the sets of data are alike. Traits unique to each set of data would be in the part of the circles that do not overlap.

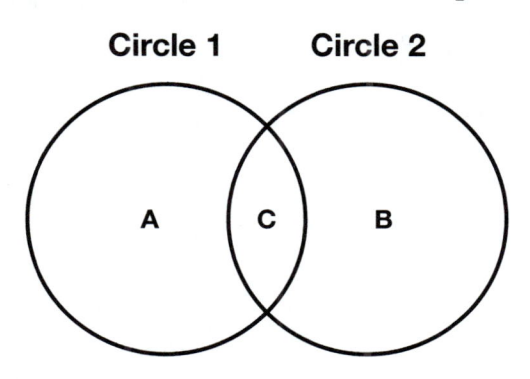

Circle 1   Circle 2

A   C   B

A represents data in Circle 1, but not in Circle 2. B represents data in Circle 2, but not in Circle 1. C represents data in both Circles 1 and 2.

A Venn diagram that compares three sets of data would look like the one shown below. Note that the ways in which A and B are alike would be listed where circles A and B overlap. The ways in which A and C are alike would be listed where circles A and C overlap, and the ways in which B and C are alike would be listed where circles B and C overlap. The ways in which A, B, and C are alike would be listed in the small area where all three circles overlap.

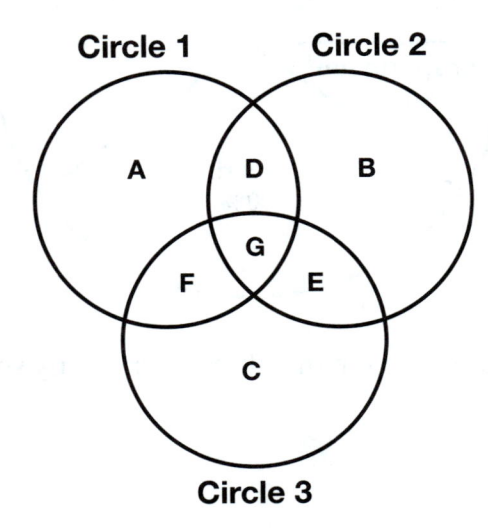

Circle 1   Circle 2

A D B

F G E

C

Circle 3

A represents data found only in Circle 1. B represents data found in only Circle 2. C represents data found in only Circle 3. D represents data found in Circles 1 and 2, but not 3. E represents data found in only Circles 2 and 3, but not 1. F represents data found in Circles 1 and 3, but not 2. G represents data found in all 3 Circles.

Let's Review 15: Graphs

Complete each of the following questions. Use the Tip below each question to help you choose the correct answer. When you finish, check your answers with those at the end of Chapter 8.

1 Which conclusion about the following graph is true?

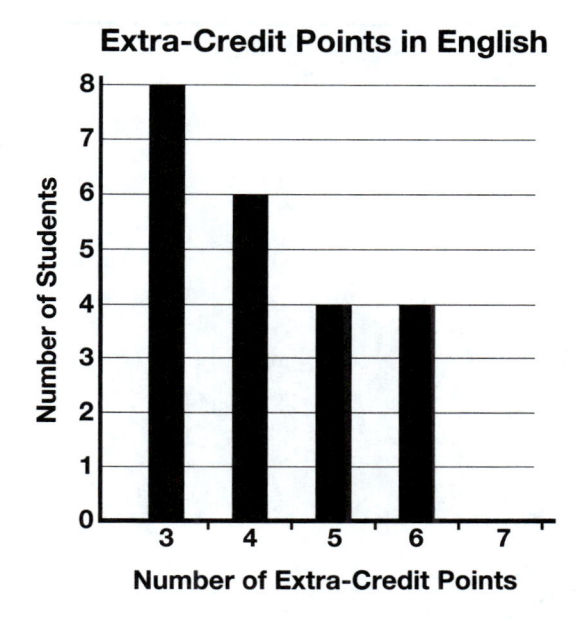

Extra-Credit Points in English

[Bar graph showing Number of Students vs Number of Extra-Credit Points: 3 points = 8 students, 4 points = 6 students, 5 points = 4 students, 6 points = 4 students]

A   More students received 4 extra-credit points than 6 extra-credit points.

B   Fewer students received 3 extra-credit points than 5 extra-credit points.

C   More students received 6 extra-credit points than 4 extra-credit points.

D   Fewer students received 4 extra-credit points than 5 extra-credit points.

TIP

Look carefully at the graph. How many extra-credit points were awarded most often? How many were awarded least often?

2 Alanis surveyed the students in her school to see what they like to do in their spare time.

Favorite Spare-Time Activity

Type	Number of Votes
Music	90
Play Sports	40
Read	44
Other	26

Which graph best represents the results of the survey?

F

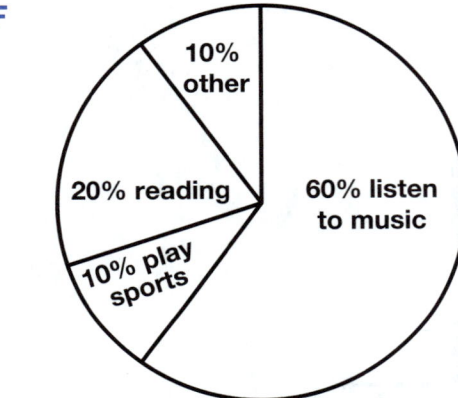

- 60% listen to music
- 10% play sports
- 20% reading
- 10% other

G

- 35% listen to music
- 20% play sports
- 25% reading
- 15% other

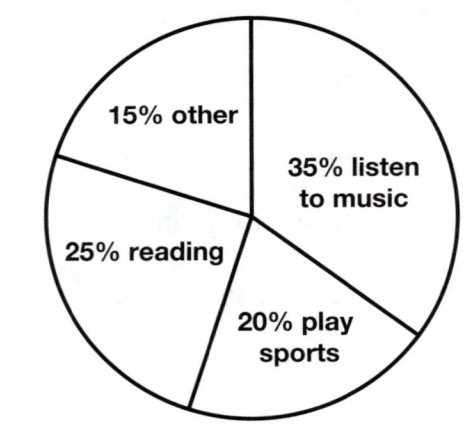

H

- 25% other
- 40% listen to music
- 20% reading
- 15% play sports

J

- 13% other
- 45% listen to music
- 22% reading
- 20% play sports

TIP: Begin by converting the number of votes into a percentage of the total.

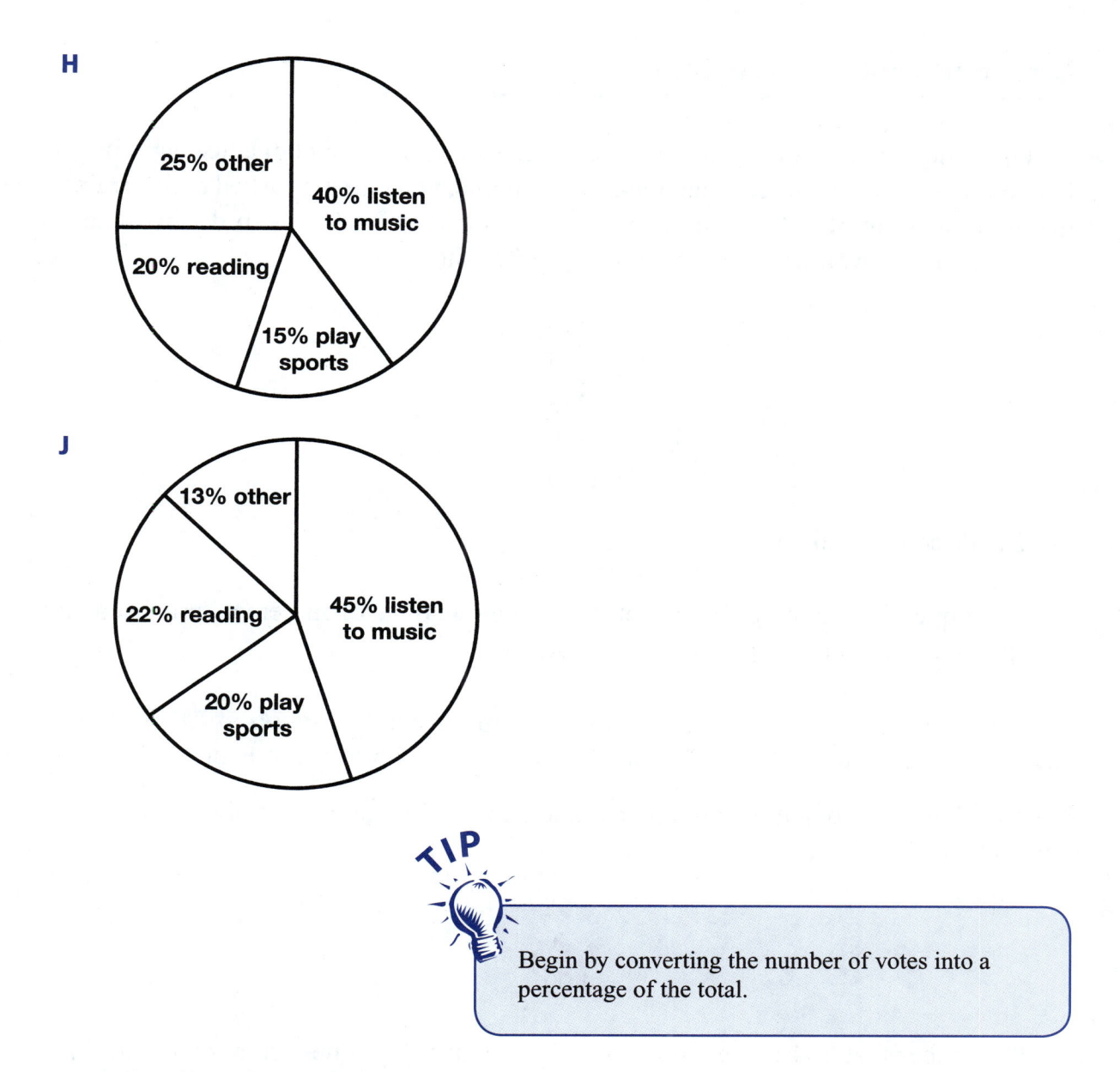

Real-Life Situations

Some questions on the TAKS will be about adding, subtracting, dividing, and multiplying monetary amounts. These questions are about real-life situations, such as calculating the sale price of a discounted item, sales tax, or the interest on a short-term loan.

Discounts and Sale Prices

For some test questions, you will have to determine the amount of a discount for an item on sale or the sale price of an item. For example, you might be asked to determine the dollar amount of a discount of 15% on shoes that cost $45.00. To do this with a paper and pencil, you would multiply 45 by .15 as shown here:

$$\begin{array}{r} 45 \\ \times .15 \\ \hline 225 \\ +450 \\ \hline 6.75 \end{array}$$

The discount is $6.75.

Other questions may ask you to determine the sale price of an item after a discount is applied. For example, read the problem below:

Megan wants to buy a mirror for her room that is usually priced at $85.00 and is now discounted by 40%. What is the sale price of the mirror?

To solve this problem using a pencil and paper, you would multiply .40 by 85 as shown here:

$$\begin{array}{r} 85 \\ \times .40 \\ \hline 34.00 \end{array}$$

Remember that $34 is the amount of the discount. This question asks you to find the sale price of the mirror, so you have to subtract 34 from 85:

$$\begin{array}{r} 85.00 \\ -34.00 \\ \hline 51.00 \end{array}$$

The sale price of the mirror is $51.00.

You determine the sales tax on an item in much the same way that you determine a discount. As an example, read this question:

Gail works in a small hardware store where the cash register does not compute the sales tax. If the sales tax is 7%, what amount should Gail add to a purchase of $10.00?

To answer this question, multiply 10 by 7%, or .07. The amount of sales tax Gail should add to a purchase of $10.00 is $0.70 or 70 cents.

Some questions might ask you to add the sales tax to the cost of item, as in this problem:

Brian wants to buy a bike that costs $125. He knows that he will have to pay 6% sales tax on the bike. How much money, including tax, does Brian need to buy the bike?

To answer this question, you have to calculate the sales tax and add it to the cost of the bike. Multiply 125 by 6%, or .06. When you do this, you get $7.50. Now add this amount onto $125, the cost of the bike. The answer is $132.50. This is the amount of money Brian needs to buy the bike.

Interest

When you borrow money, you take a loan. Usually, you're asked to pay interest on the loan. **Interest** is a sum of money you must pay in addition to the **principal**, the amount of money you borrowed. Interest is like a fee that you pay to the person or company that loaned you the money. For example, read this problem:

If Alberto borrows $5,000 from a bank at a fixed interest rate of 12% per year, how much interest must he pay if he pays the loan in full at the end of one year?

To solve this problem, you must multiply 5,000 by 12%, or .12. The answer is $600. If Alberto pays the loan in full at the end of one year, he must pay $600 in interest.

Let's Review 16: Discounts and Sale Prices

Complete each of the following questions. Use the Tip below each question to help you choose the correct answer. When you finish, check your answers with those at the end of Chapter 8.

1 Mario wants to buy a skateboard that is regularly priced at $55 but is now discounted by 15%. What is the sale price of the skateboard?

A $8.25

B $46.75

C $54.75

D $32.50

TIP
Find the amount of the discount and then deduct this amount from the price of the skateboard.

Chapter 8: Data Analysis and Probability

2 Javier works in an ice cream store where the cash register does not compute the sales tax. If the sales tax is 5%, what is the amount Javier should add to a purchase of $11.00?

> **TIP**
> Multiply 11 by 5%, or .05.

3 If a pair of jeans originally cost $25 and are selling at a 12% discount, what is the amount of this discount?

A $3.00

B $4.00

C $22.00

D $28.00

> **TIP**
> Multiply $25 by 12%, or .12, to get the amount of the discount.

4 Leo's dental plan pays 45% of dental expenses after the deductible of $100 is subtracted. Leo's total dental bill was $380. What is the exact amount the insurance company will pay?

F $109

G $126

H $226

J $280

> **TIP**
> The insurance will pay 45% of the amount after the deductible is subtracted.

5 If Javier borrows $7,000 to buy a car at a fixed interest rate of 13% per year, how much interest must he pay if he pays the loan in full at the end of two years?

A $910

B $920

C $1,820

D $1,840

TIP Notice that this question asks you to determine the interest for two years. First find the interest for one year, and then multiply this number by 2.

6 For every 90 days Kendra works, she gets 2 days sick leave and twice as many vacation days. How many vacation days does Kendra get for working 45 days?

F 1

G 2

H 4

J 6

TIP Read this problem through very carefully. You're looking for the number of vacation days, not the number of sick days.

Chapter 8 Practice Problems

Complete each of the following practice problems. Check your answers at the end of this chapter. Be sure to read the answer explanations!

1 There are 12 coins in a box; some are nickels and the rest are pennies. The probability of randomly reaching into the box and pulling out a nickel is $\frac{2}{3}$. How many pennies are in the box?

A 1

B 3

C 4

D 8

2 The weekly salaries of seven employees are $160, $240, $260, $85, $200, $180, and $120. What is the median salary?

F $120

G $160

H $180

J $200

3 If all central angles are equal, find the probability of spinning "4" on the spinner below.

A 12.5%

B 25%

C 33%

D 50%

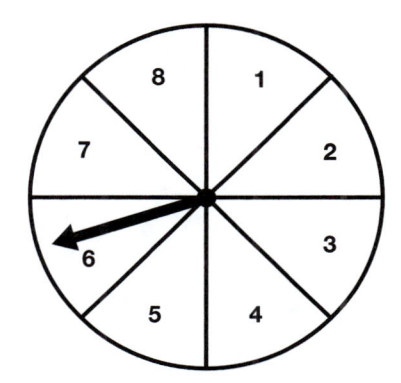

4 The average annual temperature in Sarasota, Florida, is shown in the table below.

Average Monthly High Temperature

Month	Average High Temperature
January	72°
February	74°
March	78°
April	82°
May	87°
June	91°
July	91°
August	91°
September	90°
October	85°
November	79°
December	74°

Which measure gives the highest temperature?

F Mean

G Median

H Mode

J Range

5 Christine has a bag of 25 marbles. Five of these marbles are green, 4 are blue, 3 are white, 8 are black, and 5 are yellow. If Christine reaches into the bag without looking, what is the probability that she will randomly pull out a yellow marble?

A $\frac{1}{8}$

B $\frac{1}{5}$

C $\frac{1}{4}$

D $\frac{1}{3}$

6 Kate works in a small gift shop where the cash register does not compute the sales tax. If the sales tax is 6%, how much should Kate charge the customer for a purchase of $25.00?

F $1.50

G $1.75

H $26.50

J $41.50

7 If Karen borrows $8,000 from a bank at a fixed interest rate of 14% per year, how much interest must she pay if she pays the loan in full at the end of one year?

A $1,120

B $2,120

C $5,880

D $6,880

Chapter 8 Answer Explanations

Let's Review 13: Probability

1. C

The spinner is divided into six equal sections and two of these sections are green. So the probability that the spinner will land on green is $\frac{2}{6}$, or $\frac{1}{3}$.

2. F

The bag contains 20 chips altogether and 5 of these chips are red. The probability of pulling out a red chip is therefore $\frac{5}{20}$, or $\frac{1}{4}$. This fraction can be converted to 25%.

3. D

If the number cube has six sides and is numbered 1, 2, 3, 4, 5, and 6, three of the six sides are even. Therefore, the probability of rolling an even number is $\frac{3}{6}$, or $\frac{1}{2}$.

4. J

The percentage of pink jelly beans is 25%, so the rest (75%) must be yellow.

5. C

If you toss a penny, the probability of it coming up heads or tails is $\frac{1}{2}$. The probability is always $\frac{1}{2}$ regardless of how many times you have already tossed the penny.

Let's Review 14: Central Tendency

1. B

To solve this problem, you need to calculate each measure of central tendency and the range, and then choose the measure showing the greatest number of calls. The mean is 9.5, the mode is 12, the median is 9 (the average of the middle numbers 8 and 10), and the range is 11. Therefore, the mode shows the greatest number of calls.

2. H

To find the range in the number of cars sold from January to May, subtract the smallest number, 12, from the greatest number, 58. The range is 46.

3. B

When you add all of the numbers, you get 372. When you divide 372 by the number of items, 12, you get 31, the correct answer for the mean.

4. J

In order to determine that the sales varied by $250 during the week, the manager must subtract the lowest number of sales from the highest number of sales. This would help her to determine the range of sales. J is the correct answer.

5. 88

If you add all of Renee's grades together, you get a total of 704. Divided by 8, her mean grade would be 88.

Let's Review 15: Graphs

1. A
To solve this problem, look closely at the graph. The only answer choice that presents a correct conclusion is answer choice A: Six students received 4 extra-credit points and only four students received 6 extra-credit points.

2. J
For this problem, you have to choose the circle graph displaying the correct information. Before you can do this, you need to change the number of votes for each activity into a percent. When you do this, the correct percentages are: 45% listen to music; 20% play sports; 22% reading; 13% other.

Let's Review 16: Discounts and Sale Prices

1. B
When you multiply $55 by .15, you get $8.25. When you subtract this amount from the original cost of the skateboard, $55, the answer is $46.75.

2. $0.55
To find the amount of the sales tax, you need to multiply $11 by .05. The answer is $0.55.

3. A
To solve this problem, you have to multiply $25 by 12%, or .12. The amount of the discount is $3.00.

4. G
The first step in solving this problem is to subtract $100 from $380. Then multiply this difference by 45%, or .45. The answer is $126. This is the amount Leo's insurance will pay.

Chapter 8: Data Analysis and Probability

5. C

To solve this problem you have to multiply $7,000 by 13%, or .13. The answer is $910. However, since you need to find the amount of interest Javier would pay after two years, you need to multiply $910 by 2 to get $1,820.

6. G

The correct answer is G. For every 90 days of work, Kendra gets 4 vacation days. Thus, for working 45 days, which is half of 90, Kendra should get $\left(\frac{1}{2}\right)(4) = 2$ vacation days.

Chapter 8 Practice Problems

1. C

You know that the denominator must be 12 (the number of coins), so the probability of pulling out a nickel before you reduce the fraction is $\frac{2}{3} = \frac{8}{12}$. Therefore, there are 8 nickels, and the rest are pennies, or 12 – 8 = 4 pennies.

2. H

If you put the salaries in order from least to greatest, you'll see that $180 is the middle number, or median salary.

3. A

The spinner has eight sections and only one section is numbered "4," so the probability of spinning a 4 is $\frac{1}{8}$, or 12.5%.

4. H

The mean is $\left(\frac{994}{12}\right) = 82.8°$; the median is $\frac{(82 + 85)}{2}$ 83.5°; the mode is 91°; and the range is (91 – 72) = 19 degrees. Therefore, the mode gives the highest temperature.

5. B

There are 25 marbles in the bag and 5 of them are yellow. Therefore, the probability of choosing a yellow marble is $\frac{5}{25}$, or $\frac{1}{5}$.

6. H

To solve this problem, you need to determine 6% of $25 and then add this amount onto the purchase price of $25:

$25 \times .06 = \$1.50$

$\$25.00 + \$1.50 = \$26.50$

7. A

For this problem, you need to determine 14% of $8,000:

$\$8,000 \times .14 = \$1,120$

Chapter 9
Problem Solving

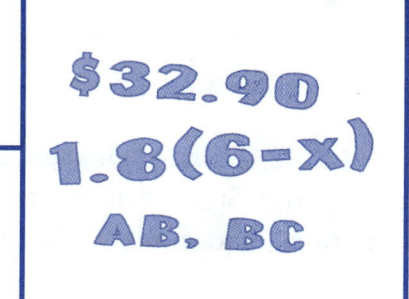

$32.90
1.8(6-x)
AB, BC

Objectives

Objective 10: The student will demonstrate an understanding of the mathematical processes and tools used in problem solving.

    (8.14) **Underlying processes and mathematical tools.** The student applies Grade 8 mathematics to solve problems connected to everyday experiences, investigations into other disciplines, and activities in and outside of school. The student is expected to

    (A) identify and apply mathematics to everyday experiences, to activities in and outside of school, with other disciplines, and with other mathematical topics;

    (B) use a problem-solving model that incorporates understanding the problem, making a plan, carrying out the plan, and evaluating the solution for reasonableness; and

    (C) select or develop an appropriate problem-solving strategy from a variety of different types, including drawing a picture, looking for a pattern, systematic guessing and checking, acting it out, making a table, working a simpler problem, or working backwards to solve a problem.

(8.15) **Underlying processes and mathematical tools.** The student communicates about Grade 8 mathematics through informal and mathematical language, representations, and models. The student is expected to

(A) communicate mathematical ideas using language, efficient tools, appropriate units, and graphical, numerical, physical, or algebraic mathematical models.

(8.16) **Underlying processes and mathematical tools.** The student uses logical reasoning to make conjectures and verify conclusions. The student is expected to

(A) make conjectures from patterns or sets of examples and nonexamples; and

(B) validate his/her conclusions using mathematical properties and relationships.

To answer most of the problem-solving questions on the TAKS, you will have to apply the skills you have learned in the previous chapters of this book. Some of the problems in this chapter will contain charts and graphs and most will be about everyday experiences. To solve some, you will have to think carefully and use logic. The following guidelines will help you correctly answer problem-solving questions:

- Understand the problem. Read the problem carefully and make sure you understand what it is asking you to do. Sometimes it helps to rewrite the problem into your own words.

- Sometimes it will be helpful to make a labeled drawing.

- Decide what you need to do to solve the problem. Some problems may contain more information than you need. Some might ask you what additional information is needed to solve the problem.

- Determine what mathematical operations you need to perform to solve the problem. For example, you might need to determine the square root of a number or find a percent. Turn back to earlier chapters in this book if you're not sure how to perform a mathematical operation.

- Solve the problem and check your work.

Let's Review 17: Problem Solving

Complete each of the following questions. Use the Tip below each question to help you choose the correct answer. When you finish, check your answers with those at the end of Chapter 9.

1 A conservationist has been observing spider mites on spruce trees in her state. She recorded that 15 of the 75 spruce trees she observed throughout the state appeared to be infected with spider mites. The state has an estimated 200,000 spruce trees.

Based on her observations, approximately how many spruce trees would she predict are infected with spider mites?

A 15,000

B 20,000

C 40,000

D 100,000

TIP
You need to create a proportion to solve this problem. Begin with the ratio $\frac{15}{75}$, then equate it to the number of infected trees, x, out of the total number of trees.

2 The area of Ohio is about 4.1×10^4 square miles. The area of Massachusetts is 8.0×10^3 square miles. What is the difference between the area of Ohio and the area of Massachusetts?

F 3.3×10^1

G 3.3×10^7

H 3.3×10^2

J 3.3×10^4

TIP
To answer this question, you need to convert the square miles for each state into standard form. Then subtract the area of Massachusetts from the area of Ohio and convert this number into scientific notation.

3 A party planner determines that for every 12 guests who attend the party, she will need 8 pounds of chicken. How many pounds of chicken will she need if 72 guests attend the party?

A 12

B 36

C 48

D 576

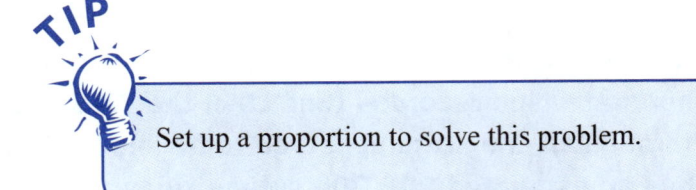

TIP Set up a proportion to solve this problem.

4 Melissa asked her class what their favorite kind of music was. She computed the data into percentages and graphed it as shown in the chart below.

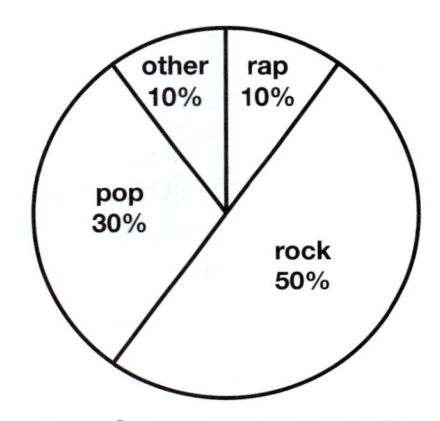

Students' Favorite Kind of Music

- other 10%
- rap 10%
- pop 30%
- rock 50%

What percentage of students chose rock as their favorite kind of music?

F 10%

G 20%

H 30%

J 50%

TIP Look at each division in the pie chart. The entire pie chart represents 100%.

5 A sneaker company surveyed 1,000 runners to find out if they liked to wear Running Shoe A or Running Shoe B. The diagram below shows the results of the survey.

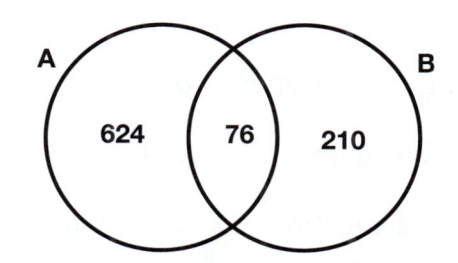

[Venn diagram with circle A containing 624, overlap containing 76, and circle B containing 210]

Which expression can be used to determine the number of runners surveyed who did not like either Running Shoe A or Running Shoe B?

A 1,000 − (624 + 210 + 76)

B 1,000 − (624 + 210)

C 624 + 210 + 76

D 624 + 210

TIP

Look carefully at the information given in this problem. Remember that in a Venn diagram, the overlapping portion of the circles represents the runner who liked both Running Shoe A and Running Shoe B. There are 1,000 runners, and 624 liked Running Shoe A, 210 liked Running Shoe B, and 76 liked both.

6 To get to her friend's house, Melanie traveled 8 miles south and 6 miles west. Which of the following describes the method for finding the straight-line distance from Melanie's house to her friend's house?

F Use $a = 8$ and $b = 6$ in the equation $c^2 = a^2 + b^2$ and then solve for c.

G Use $c = 8$ and $a = 6$ in the equation $c^2 = a^2 + b^2$ and then solve for c.

H Use $c = 8$ and $a = 6$ in the equation $c^2 = a^2 + b^2$ and then solve for a, b, and c.

J Use $a = 8$ and $c = 6$ in the equation $c^2 = a^2 + b^2$ and then solve for a, b, and c.

> **TIP**
> You need to correctly use the Pythagorean theorem to solve this problem. If Melanie traveled south and west, the straight-line distance would be the hypotenuse of the triangle.

7 If $\overline{AB}$ and $\overline{CD}$ are sides of a rhombus, which of the following must be true?

A $\overline{BC}$ and $\overline{AD}$ are parallel.

B $\overline{AB}$ and $\overline{BC}$ are parallel.

C $\overline{BC}$ and $\overline{AD}$ are perpendicular.

D $\overline{AB}$ and $\overline{CD}$ are perpendicular.

> **TIP**
> Draw a rhombus with $\overline{AB}$ and $\overline{CD}$ as sides. Then choose the correct answer.

Chapter 9: Problem Solving

8 Keisha saves 10% of her total gross weekly earnings from two part-time jobs. She earns $5.00 per hour for one part-time job and $6.25 per hour from the other part-time job. Keisha works a total of 20 hours between the two jobs each week. What additional information is needed to determine the amount of earnings she saves each week?

F The number of hours she works each month.

G The number of days she works at each job.

H The number of hours she works each day.

J The number of hours she works at each job.

> **TIP**
> You first have to figure out the total earnings Keisha makes each month. What information would you need to determine that amount?

Chapter 9 Practice Problems

1 Points *C* and *D* lie on circle *G*. If circle *G* has a radius *r*, which of the following statements cannot be true?

A $CD = 2r$

B $CD > r$

C $CD = r$

D $CD > 2r$

2 Bethany constructed a Venn diagram to illustrate the number of students in her class who have a pet.

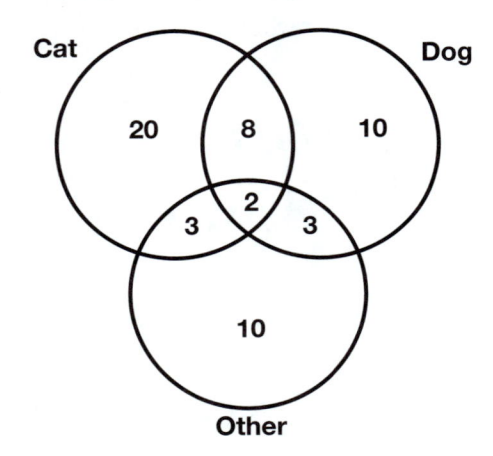

How many students have a dog and a cat and another type of pet?

F 2

G 3

H 8

J 40

3 Rachel works as a waitress. She earns $6.00 an hour plus tips. If she averages 5 customers an hour and earns $148 in an eight-hour shift, which amount best represents the average tip per customer?

A $2.50

B $5.00

C $20

D $100

4 Teresa plans to set up a lemonade stand at a local fair. She will purchase 250 cans of lemonade for $75 and will charge $2.50 for each can she sells. In addition to the cost of the lemonade, Teresa will need to pay $10 to set up the stand. Which of the following expressions could Teresa use to find out how much money she could make after expenses, for selling x cans of juice?

F $2.5x - 75 - 10$

G $x + 2.50 - 75 - 10$

H $2.50 - 75 - \dfrac{10}{x}$

J $2.5x(75 - 10)$

5 Ashaki joined her school's cross country team. As part of her training, she is going to increase the number of miles she runs every week by 2 miles. If she runs 21 miles the first week, how many miles will she run during the eighth week? Show your work below.

6 A clothing store marked all sweaters $\frac{1}{4}$ off the original price for a sale. Alex has a store coupon that is good for an additional discount of 10% off the sale price. She purchases a sweater that was originally priced at $45.00. If she uses her discount coupon, what should be the cost of the sweater before the sales tax is added?

F $14.55

G $23.75

H $30.37

J $33.75

7 Samantha and Alex ran a 2-mile race. If *s* represents the number of minutes Samantha took to finish the race and *a* represents the number of minutes Alex took to finish the race, which of the following describes a situation in which Samantha finished the race before Alex?

A $s < a$

B $s > a$

C $s \leq a$

D $s \geq a$

Chapter 9 Answer Explanations

Let's Review 17: Problem Solving

1. C

To solve this problem, reduce the fraction $\frac{15}{75}$ to $\frac{1}{5}$. Then set up a proportion like this:

$\frac{1}{5} = \frac{x}{200,000}$. Then cross-multiply:

$5x = 200,000$

$\frac{5x}{5} = \frac{200,000}{5}$

$x = 40,000$

2. J

Begin by converting the square miles for each state into standard form:

$4.1 \times 10^4 = 41,000$

$8.0 \times 10^3 = 8,000$

Then subtract: $41,000 - 8,000 = 33,000$.

Now convert 33,000 to scientific notation: 3.3×10^4.

Chapter 9: Problem Solving 213

3. C

You have to set up a proportion to solve this problem:

$$\frac{8}{12} = \frac{x}{72}$$

$$\frac{2}{3} = \frac{x}{72}$$

$$3x = 144$$

$$\frac{3x}{3} = \frac{144}{3}$$

$$x = 48$$

4. J

50% of the student chose rock as their favorite type of music.

5. A

One thousand runners were surveyed, but only 624 chose Running Shoe A, 210 liked Running Shoe B, and 76 liked both. Therefore, only 910 runners chose A or B or both. You would use the equation 1,000 − (624 + 210 + 76) to determine how may liked neither.

6. F

The distance Melanie traveled south (8) and west (6) form the legs of a right triangle, where $a = 8$ and $b = 6$. You would use $c^2 = a^2 + b^2$ to find the straight-line distance, c, which is the hypotenuse of the right triangle.

7. A

Opposite sides of a rhombus are parallel and $\overline{BC}$ and $\overline{AD}$ are opposite sides.

8. J

You would need to know how many hours she works at each job.

Chapter 9 Practice Problems

1. D

All points on a circle are an equal distance from the center. Therefore, the distance between these points can't be greater than 2 times the radius, but they may equal this amount.

2. F

You need to look at the part where all three circles overlap to solve this problem. Two students have a dog and a cat and another type of pet.

3. A

To solve this problem, you need to first determine how much Rachel earns in an eight-hour shift without tips. $6.00 \times 8 = $48. To find her tip income, deduct this amount from $140: $148 − $48 = $100. She earned $100 in tips in an eight-hour shift. To determine her average tip per customer, if she averages 5 customers per hour, she averages about 40 customers in 8 hours, or the average tip per customer is $100 ÷ 40 = $2.50.

4. F

The variable in this problem is x, the number of cans of lemonade Teresa sells. Teresa will charge $2.50 for each can, so she brings in $2.5x$. From this amount, she must subtract her expenses of $75 and $10. So the best expression is $2.5x - 75 - 10$, answer choice F.

5. 35

The number of miles Ashaki will run during the 8th week is $21 + (2)(7) = 35$.

6. H

First deduct 25% from the original cost of the sweater, $45: 45 × .25 = $11.25. $45 − $11.25 = $33.75. Then take an additional 10% off the price of the sweater:

$33.75 × .10 = $3.38
$33.75 − $3.38 = $30.37

7. A

Samantha finished the race in less time than Alex, so the correct equation is $s < a$.

TAKS Mathematics Practice Test

1

Directions: This Practice Test contains 60 questions. A graphing calculator may be used.

Mark answers in the Answer Document at the end of this test.

Mathematics Chart – Measurements

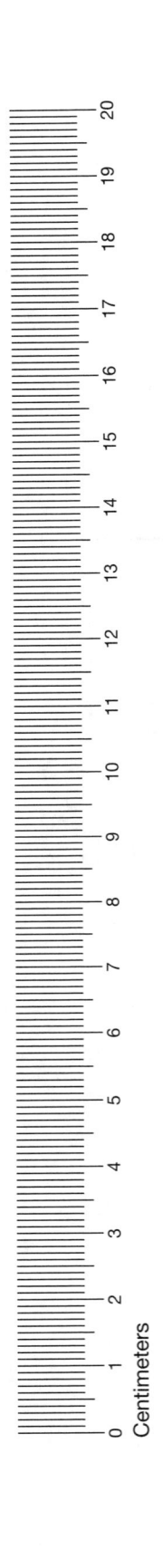

LENGTH

Metric

1 kilometer = 1000 meters

1 meter = 100 centimeters

1 centimeter = 10 millimeters

Customary

1 mile = 1760 yards

1 mile = 5280 feet

1 yard = 3 feet

1 foot = 12 inches

CAPACITY AND VOLUME

Metric

1 liter = 1000 milliliters

Customary

1 gallon = 4 quarts

1 gallon = 128 ounces

1 quart = 2 pints

1 pint = 2 cups

1 cup = 8 ounces

MASS AND WEIGHT

Metric

1 kilogram = 1000 grams

1 gram = 1000 milligrams

Customary

1 ton = 2000 pounds

1 pound = 16 ounces

TIME

1 year = 365 days

1 year = 12 months

1 year = 52 weeks

1 week = 7 days

1 day = 24 hours

1 hour = 60 minutes

1 minute = 60 seconds

Mathematics Chart – Formulas

Perimeter	rectangle	$P = 2l + 2w$ or $P = 2(l + w)$
Circumference	circle	$C = 2\pi r$
Area	rectangle	$A = lw$ or $A = bh$
	triangle	$A = \dfrac{1}{2}bh$ or $A = \dfrac{bh}{2}$
	trapezoid	$A = \dfrac{1}{2}(b_1 + b_2)h$ or $\dfrac{(b_1 + b_2)h}{2}$
	circle	$A = \pi r^2$
Surface Area	cube	$S = 6s^2$
	cylinder (lateral)	$S = 2\pi rh$
	cylinder (total)	$S = 2\pi rh + 2\pi r^2$ or $S = 2\pi r(h + r)$
	cone (lateral)	$S = \pi rl$
	cone (total)	$S = \pi rl + \pi r^2$ or $S = \pi r(l + r)$
	sphere	$S = 4\pi r^2$
Volume	prism or cylinder	$V = Bh^*$
	pyramid or cone	$V = \dfrac{1}{3}Bh^*$
	sphere	$V = \dfrac{4}{3}\pi r^3$

*B represents the area of the base of a solid figure.

Pi	π	$\pi \approx 3.14$ or $\pi \approx \dfrac{22}{7}$
Pythagorean Theorem		$a^2 + b^2 = c^2$
Distance Formula		$d = \sqrt{(x_2 - x_1)^2 + (y_2 - y_1)^2}$
Slope of a Line		$m = \dfrac{y_2 - y_1}{x_2 - x_1}$
Midpoint Formula		$M = \left(\dfrac{x_1 + x_2}{2}, \dfrac{y_1 + y_2}{2}\right)$
Quadratic Formula		$x = \dfrac{-b \pm \sqrt{b^2 - 4ac}}{2a}$
Slope-Intercept Form of an Equation		$y = mx + b$
Point-Slope Form of an Equation		$y - y_1 = m(x - x_1)$
Standard Form of an Equation		$Ax + By = C$
Simple Interest Formula		$I = prt$

1. A researcher made a table showing the relationship between the year and the number of people who have cell phones. What is the independent quantity in the relationship?

   A Year

   B People with cell phones

   C All of the data in the table

   D Cannot be determined

2. Acme Corporation pledged to reduce the amount of pollution it emits by half each year. If it emitted 512,000 tons of pollution in 2005, how much can it emit in 2012?

   F 2,000 tons

   G 4,000 tons

   H 8,000 tons

   J 16,000 tons

3. Triangle ABC has vertices at the coordinates (–5, 1), (–7, 5), and (–3, 5), as shown.

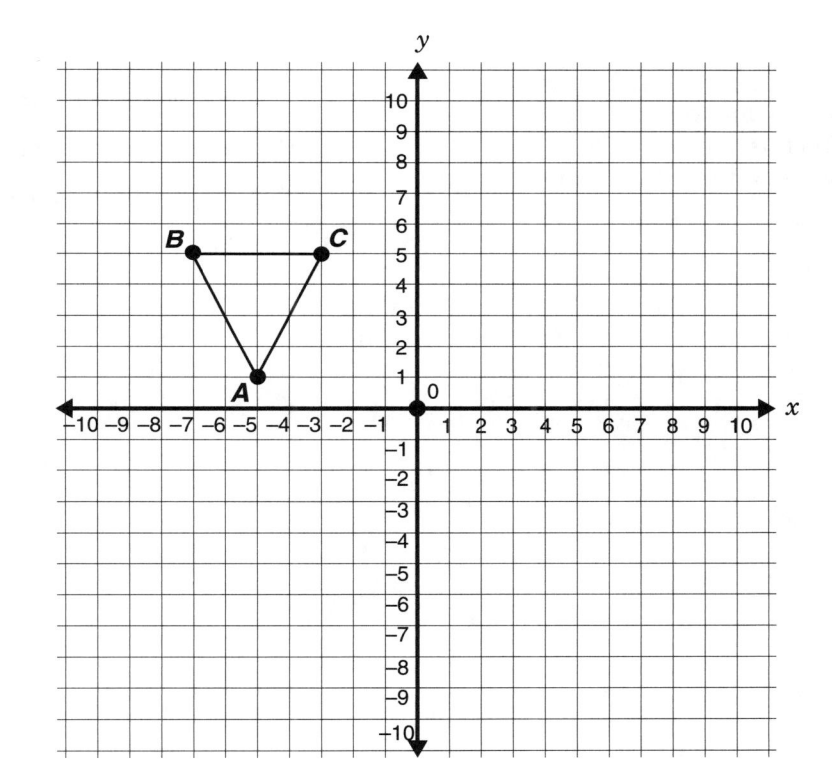

   What are the coordinates of the vertices of triangle ABC when it is reflected over the x-axis?

   A (–7, –5), (–3, –5), (–5, –1)

   B (–5, –1), (–7, –5), (–3, 5)

   C (5, 1), (3, 5), (7, 5)

   D (5, –1), (7, –5), (3, –5)

GO ON

4. A function is described by the equation $y = 3x - 1$, in which y is dependent on x. If a value for the independent variable is selected from the set $\{-4, -2, 0, 3\}$, which of the following can be a corresponding dependent value?

F −13

G −10

H 0

J 3

5. Ms. Ramirez designed a true-false English exam. The ratio of true answers to false answers is 4:3. Which is closest to the percentage of true answers on Ms. Ramirez's English exam?

A 45%

B 57%

C 80%

D 132%

6. Riverside Junior High conducted a survey of students' favorite school meals. The table below shows the results of the survey.

Favorite School Meals

Meal	Number of votes
Burritos	185
Chef Salad	92
Fish Fingers	68
Pizza	193
Spaghetti	114

Which conclusion can be drawn from the results of the survey?

F There is no clear favorite.

G Salads should be served more often.

H Fish fingers is the least favorite.

J Spaghetti is the mean result.

7. What is the apparent slope of the line graphed below?

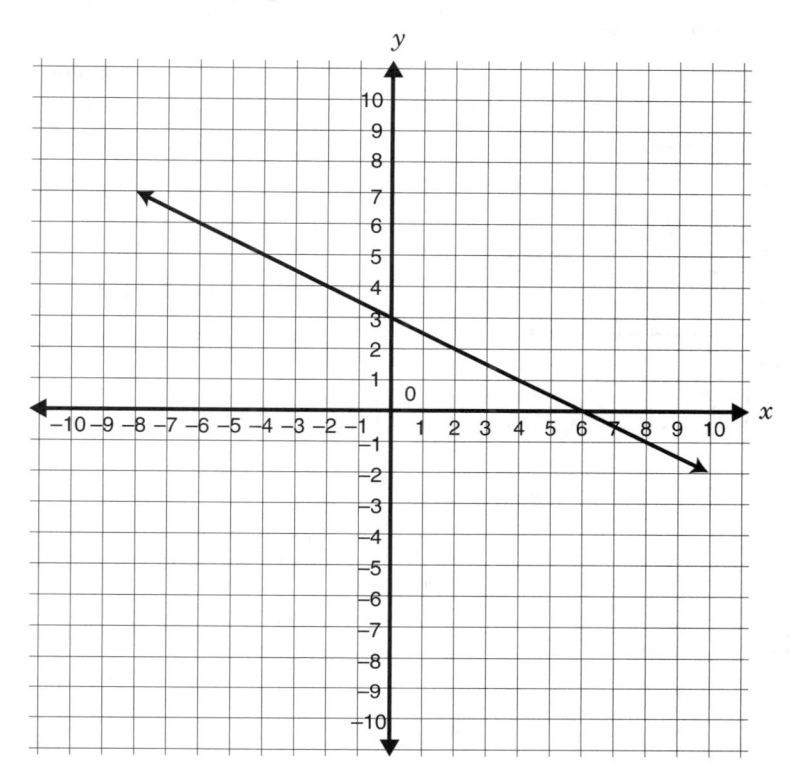

A −2

B $-\dfrac{1}{2}$

C $\dfrac{1}{2}$

D 2

8. A bag contains 8 red tiles, 12 blue tiles, 7 yellow tiles, 5 orange tiles, and 8 green tiles. If you reach into the bag without looking, what is the probability of pulling out an orange tile?

F $\dfrac{1}{10}$

G $\dfrac{1}{8}$

H $\dfrac{1}{5}$

J $\dfrac{5}{8}$

9. Which ordered pair represents one of the roots of the function $f(x) = x^2 - 5x + 6$?

A (−3, 0)

B (−2, 0)

C (2, 0)

D (6, 0)

GO ON

10. The graph of the equation $y = -2x + 8$ is given below. Graph $y = \frac{1}{3}x + 1$ on the grid.

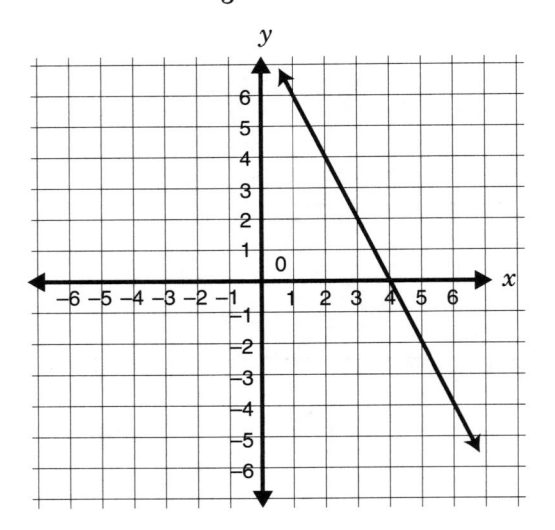

What is the solution to this system of equations?

F $(1\frac{1}{3}, 5)$

G $(-4, -3)$

H $(2, 1\frac{1}{3})$

J $(3, 2)$

11. A mail-order candle manufacturer charges $12 per candle, plus 3% shipping and handling. If s = candle sales, y = number of candles sold, and c = total charges, which of the following equations shows how to determine the total charges?

A $c = \$12.00y + .03s$

B $c = \$12.00y - .03s$

C $c = (\$12.00y)(.03s)$

D $c = (\$12.00y)(.003s)$

12. In the figure below, the shaded area is a planar cross-section of a rectangular solid.

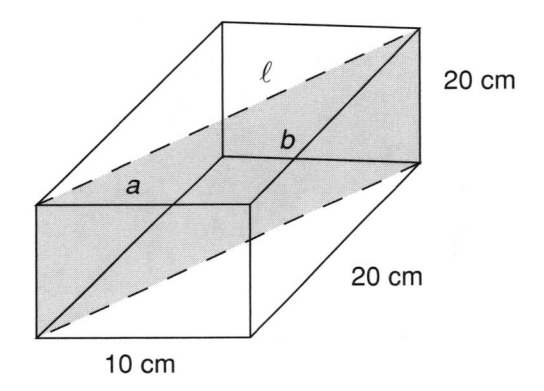

To the nearest square centimeter, what is the area of the cross section?

Record your answer and fill in the bubbles on your answer document. Be sure to use the correct place value.

GO ON

13. Which of the following graphs best represents a curve that is symmetric with respect to only the x-axis?

A

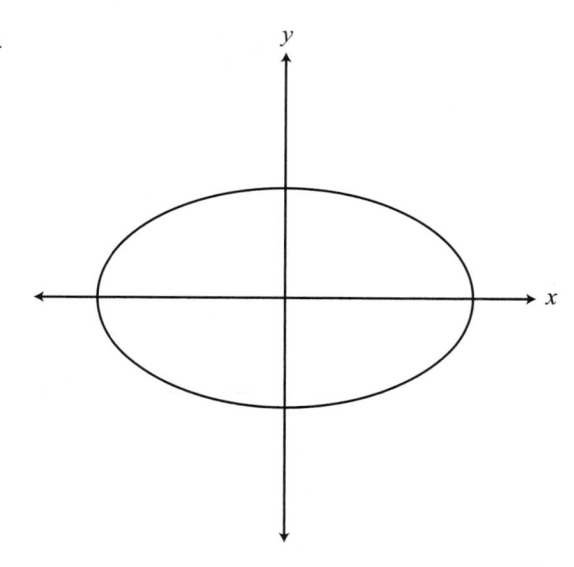

C

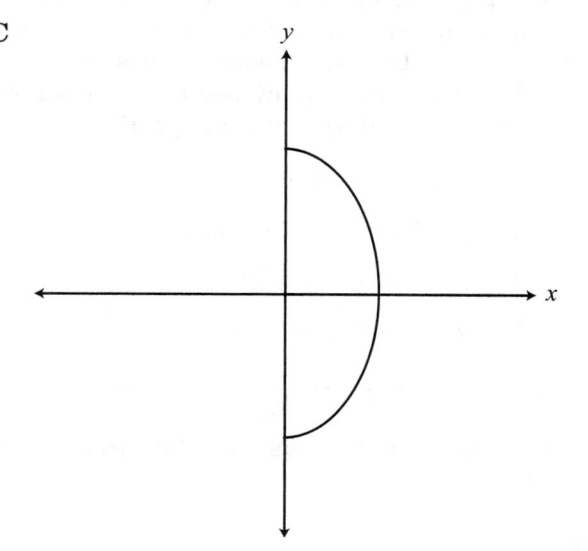

B

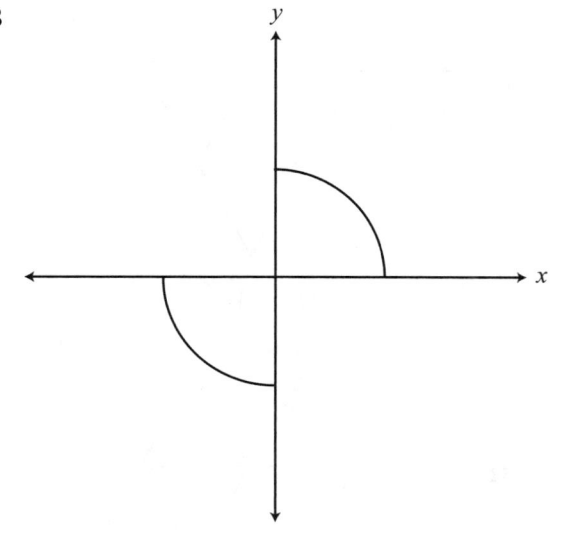

D

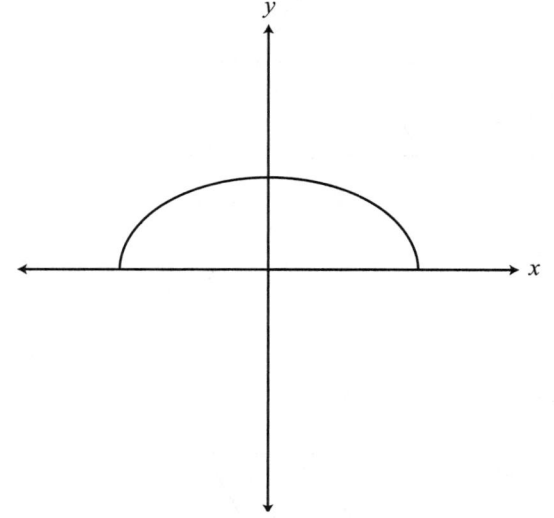

14. Sharon decided to invest some money she earned while working a summer job. She invested $4000 of the money at an annual rate of 3% and the rest of the money, x, at an annual rate of 5.50%. Which equation describes y, the total amount of interest earned from both investments during the first year?

F $y = 3(4000) + 5.5x$

G $y = (4000 + x)(0.03 + 0.550)$

H $y = (4000 + x) + 5.50x$

J $y = 0.03(4000) + 0.055x$

15. Which net represents the octahedron shown below?

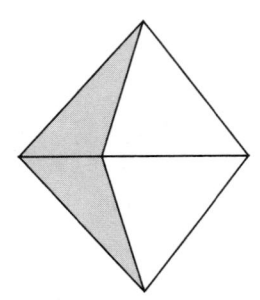

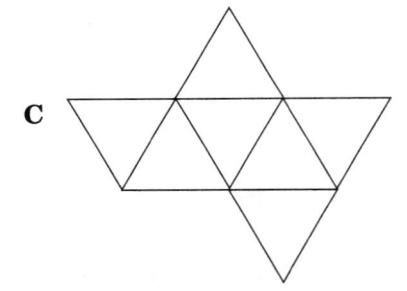

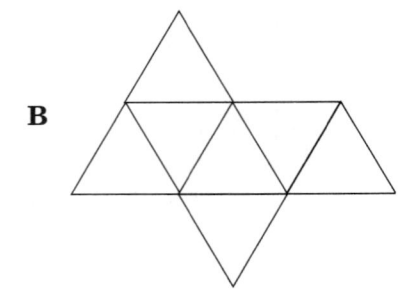

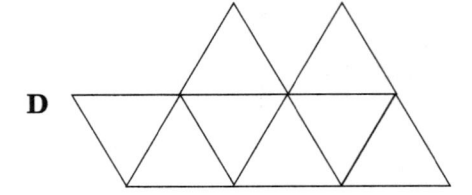

16. The graph below shows parallelogram *DEFG*, and line segments $\overline{KL}$ and $\overline{LM}$.

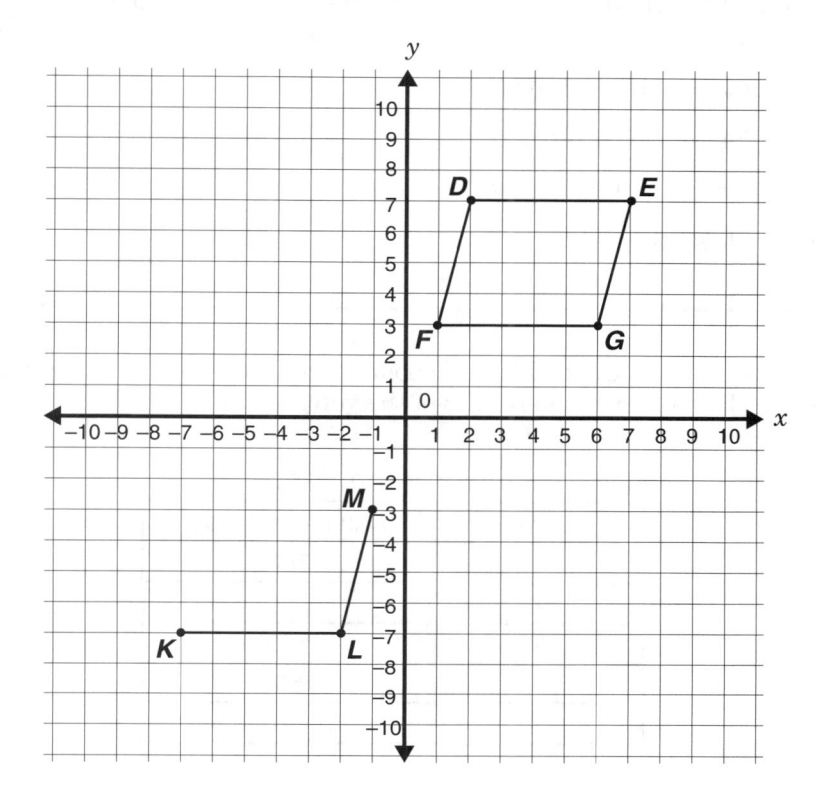

At what coordinate point should vertex *N* be placed to make parallelogram *KLMN* congruent to parallelogram *DEFG*?

F (–6, –2)

G (–6, –3)

H (–5, –3)

J (–5, –2)

17. A fitness center has a total of 85 gold and silver memberships worth $140,000. The value of a gold membership is $640 and the value of a silver membership is $400. Which system of equations can be used to find g, the number of gold memberships, and s, the number of silver memberships?

A $g + s = 85$ $640g + 400s = 140,000$

B $g + s = 85$ $74g + 400s = 140,000$

C $g + s = 45$ $400g + 640s = 140,000$

D $g - s = 140,000$ $400g + 640s = 85$

---

18. A total of 550 students from the Central High School junior class voted on their choice of a band for the junior prom. The table below shows the results of the vote.

Junior Prom Band

Band	Number of Votes
1	280
2	72
3	110
4	50
5	38

Which graph best represents the results of the survey?

F B1 50%, B2 15%, B3 25%, B4 8%, B5 2%

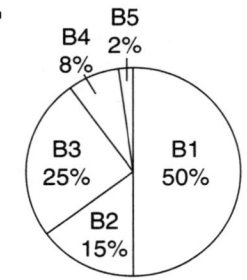

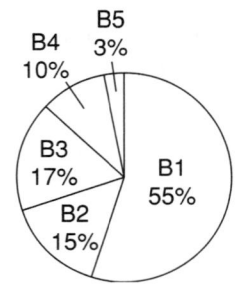

G B1 48%, B2 20%, B3 20%, B4 5%, B3 7%

H B1 55%, B2 15%, B3 17%, B4 10%, B5 3%

J B1 51%, B2 13%, B3 20%, B4 9%, B5 7%

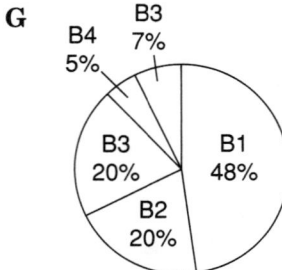

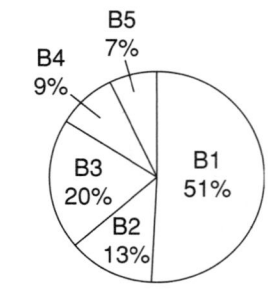

GO ON

19. Renee needs to simplify the following expression for her homework assignment.

$3(2x - y) + 4(2x + y) + 3(x + y)$

Which of the following expressions is equivalent to the expression above?

A $17x + 4y$

B $3x + 4y$

C $17x + 14y$

D $17x - 4y$

20. The polygon ABCD is shown below.

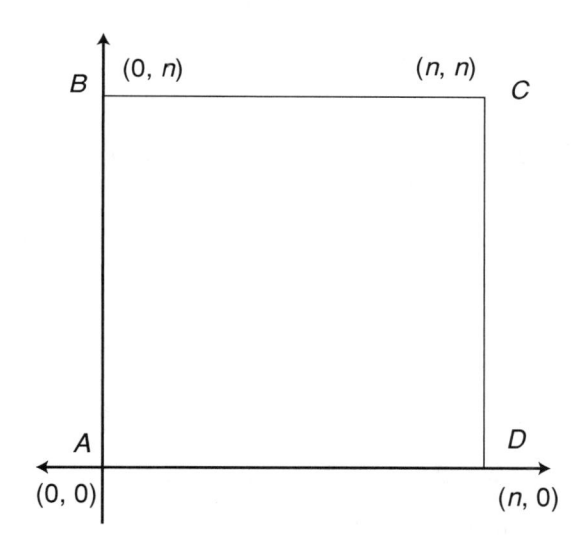

B (0, n)   (n, n) C

A (0, 0)   (n, 0) D

What expression represents the area of ABCD?

F n^2

G $2n^2$

H 0

J $4n^2$

21. Which of the following is a sufficient condition to show that a certain equation represents a linear function?

A The graph of the equation has more than one y-intercept.

B The graph of the equation has more than one x-intercept.

C The graph of the equation is a parabola.

D The graph of the equation is a non-vertical straight line.

22. The population of Dallas, Texas, is close to 1.18×10^6, and the population of Houston, Texas, is about 1.95×10^6. What is the combined population of both cities?

F 3.13×10^6

G 3.13×10^{12}

H 4.00×10^{12}

J 4.12×10^{36}

GO ON

23. The regular octagon below shows selected positions on a combination lock.

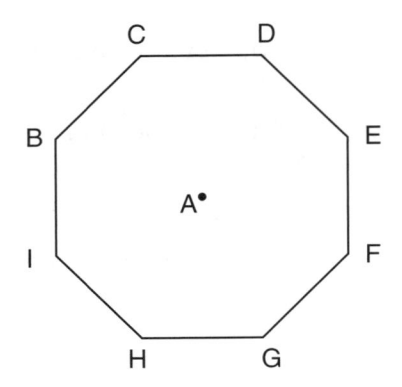

The dial of the lock is turned 90° counter-clockwise and then 225° clockwise. Which pair of points can describe the starting and ending points of a marker on the dial of this lock?

A E to B
B B to D
C I to G
D C to F

24. Triangle ABC is a right isosceles triangle. What is the measure of angle B?

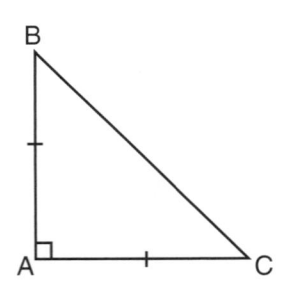

F 30°
G 45°
H 60°
J 90°

25. Shamus earns $10.00 an hour at his summer job. His employer must pay him "time and a half" (1 ½ times his regular hourly earnings) for each hour over 40 hours per week. His employer withholds 16% of his gross pay for various taxes. The table below shows Shamus's work time for the week.

Shamus's Hours

Mon	Tue	Wed	Thu	Fri
$8\frac{1}{2}$ h	10 h	$9\frac{3}{4}$ h	$8\frac{1}{4}$ h	9 h

What is Shamus's take-home pay for the week?

Record your answer and fill in the bubbles on your answer document. Be sure to use the correct place value.

26. Which of the following must be true for the isosceles trapezoid shown below?

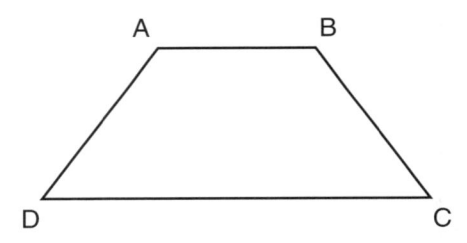

F Lines AD and BC are parallel.
G Lines BC and CD are perpendicular.
H ∠A and ∠C are congruent.
J ∠C and ∠D are congruent.

27. The scatterplot below shows the relationship between the number of cups of coffee students drink per week and their test scores. What kind of trend is shown?

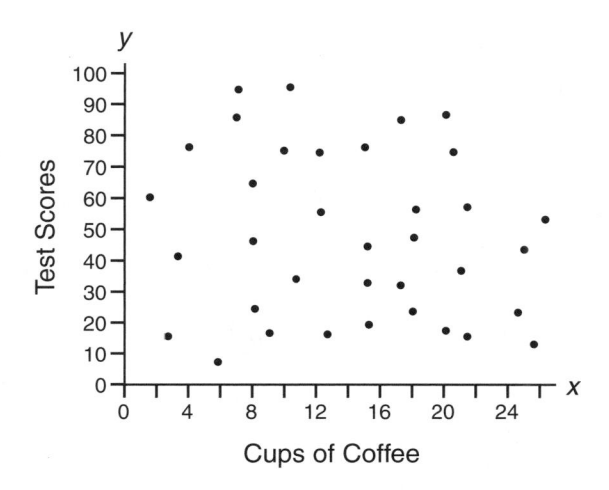

[Scatterplot: Test Scores (y-axis, 0 to 100) vs Cups of Coffee (x-axis, 0 to 24)]

A Positive trend
B Negative trend
C No trend
D Cannot be determined

28. A salesperson earns $325 per week plus a commission of $\frac{1}{6}$ of her total sales. If her sales total x dollars, which equation can be used to determine her total weekly earnings designated as y?

F $y = 325 + \frac{1}{6}x$

G $y = 325 \times \frac{1}{6}x$

H $y = \frac{1}{6}(325 + x)$

J $y = 325x + \frac{1}{6}$

29. The equation $F = \frac{9}{5}C + 32$ represents the relationship between F, the temperature in degrees Fahrenheit, and C, the temperature in degrees Celsius. If the temperature is 95°F, what is the temperature in degrees Celsius?

Record your answer and fill in the bubbles on your answer document. Be sure to use the correct place value.

GO ON

30. Rex designed a floor pattern for his room. He used only horizontal reflections of the following tile to produce the pattern.

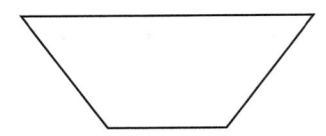

Which pattern did Rex produce?

F

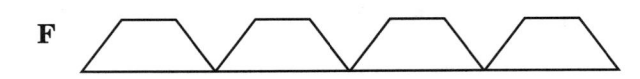

G

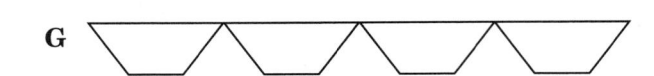

H

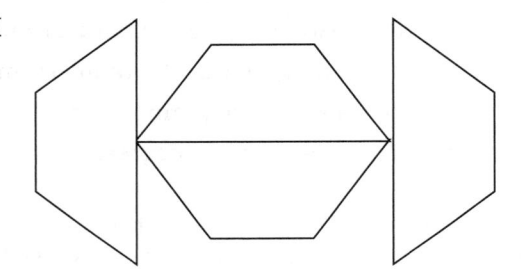

J

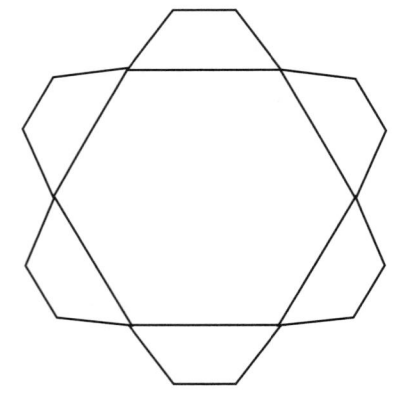

31. If the mean number of people who visited a museum over 5 days is 250, what is the total attendance during the 5 days?

A 250

B 1,000

C 1,250

D 2,500

32. Look at the figure shown below.

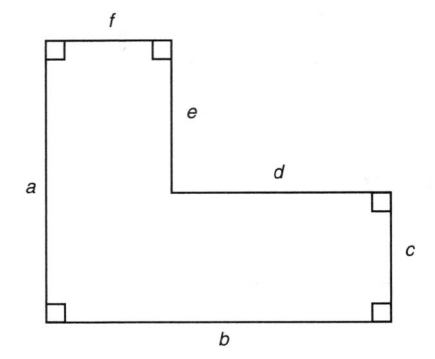

Which expression represents the area of the figure?

F $af + bc$

G $cd + bf - fe$

H $af + cd$

J $ed - bc$

GO ON

33. Look at the line shown on the coordinate grid below.

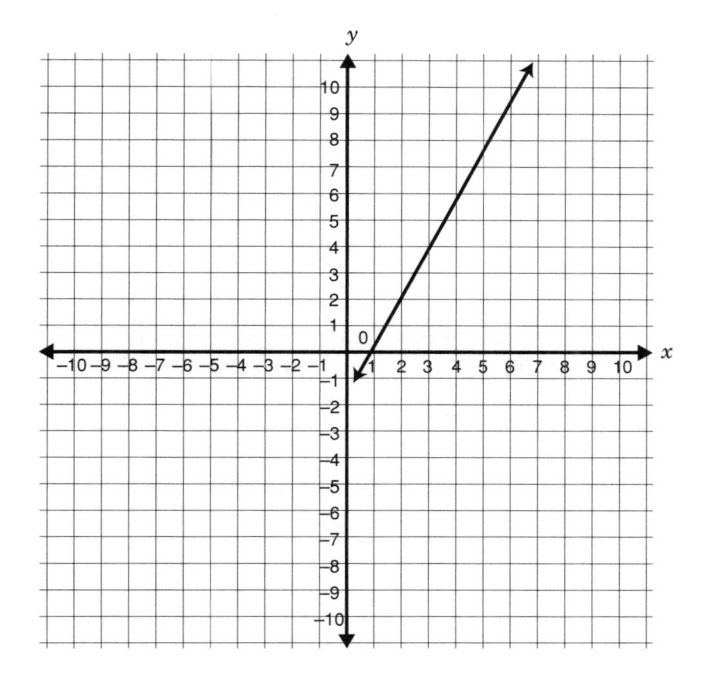

Which of the following best represents an equation of this line?

A $y = \frac{1}{2}x + 2$

B $y = 2x + 2$

C $y = 2x - 2$

D $y = \frac{1}{2}x - 2$

34. Kayla needs to make a cube for a school project. She plans to cut a shape out of thick cardboard and then fold it to make a cube. Which of these nets or shapes could Kayla use to make her model?

F

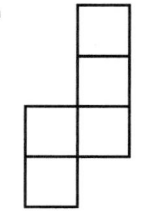

G

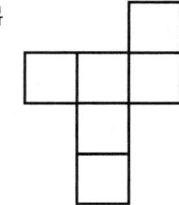

H

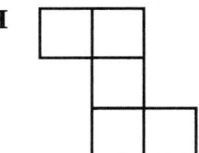

J

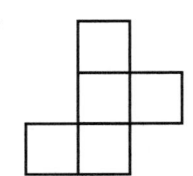

35. The graph below shows the number of students in Amy's town graduating from college from 2001 to 2005.

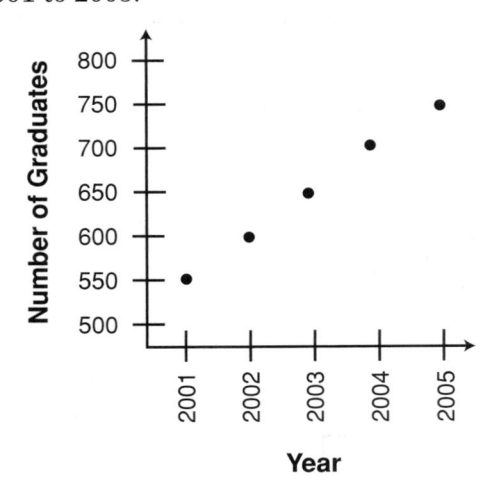

If the trend shown on the graph continues, in what year will there be 900 college graduates?

A 2006

B 2007

C 2008

D 2009

36. Line a intersects parallel lines b and c, as shown

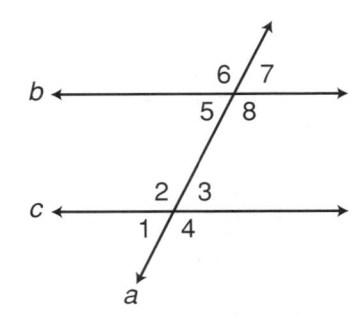

Which of the following relationships must be true, where $m\angle 5$ means "measure of $\angle 5$"?

F $m\angle 5 = m\angle 3$

G $m\angle 8 = m\angle 3$

H $m\angle 6 = m\angle 1$

J $m\angle 2 = m\angle 3$

37. Jordan is two less than three times as old as Cayce. Which expression represents this statement?

A $j = -2 + 3c$

B $j - 2 = 3c$

C $j = 3(c - 2)$

D $3j - 2 = c$

38. The figure below shows the position of a light (N) shining onto a parking lot. Angles CAD and ACE are congruent.

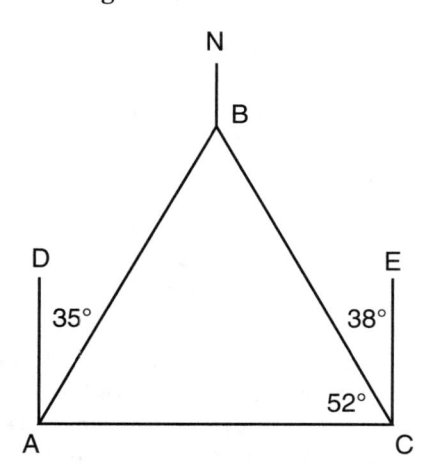

Which of the following best represents the measure of ∠CAB?

F 35°

G 38°

H 55°

J 145°

39. Kaitlyn has $80 to go shopping. She wants to buy a jacket, a, and a pair of shoes, b. Which inequality shows how much Kaitlyn can spend?

A $a + b = \$80$

B $a + b \leq \$80$

C $a + b < \$80$

D $a + b \geq \$80$

40. The graph of the equation $y = x^2$ is shown below.

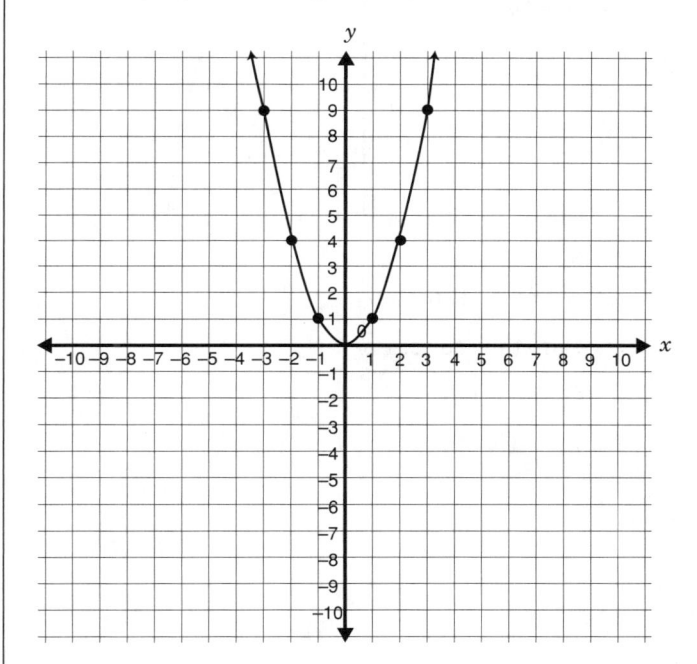

If the graph is translated 7 units down, what will be the equation of the resulting graph?

F $y = x^2 - 7$

G $y = 2x^2 - 7$

H $y = x^2/7$

J $y = x - 7$

GO ON

41. The graph below represents which type of parent function?

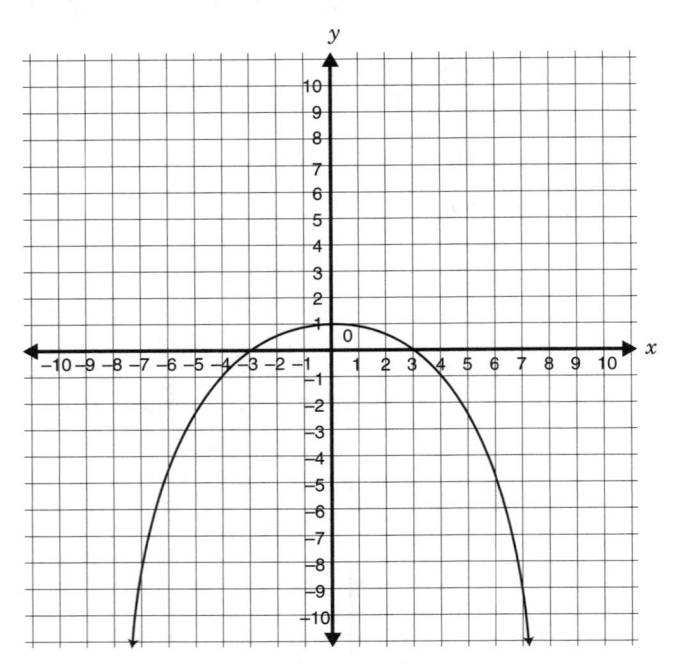

A Absolute function
B Constant function
C Quadratic
D Identity

42. Michelle is planning the finale of a dance show, in which there will be four concentric circles of dancers, as shown below.

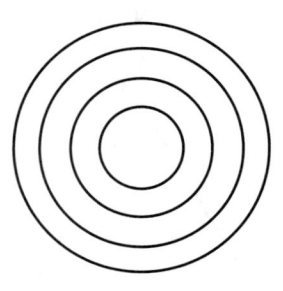

The innermost circle will consist of 12 dancers. The next larger circle will consist of 14 dancers. If this pattern of increase in the number of dancers continues for the remaining two circles, what is the total number of dancers in all four circles?

Record your answer and fill in the bubbles on your answer document. Be sure to use the correct place value.

43. What is the volume of the box pictured below?

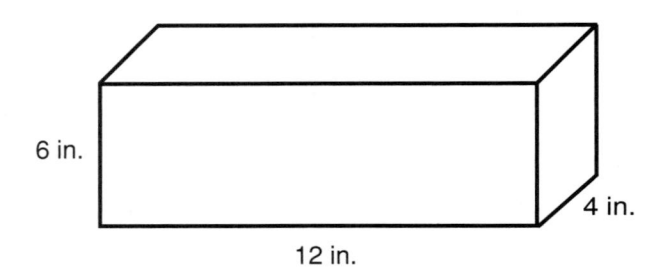

6 in.
4 in.
12 in.

A 22 in.^{3}
B 72 in.^{3}
C 288 in.^{3}
D 576 in.^{3}

44. Minh wants to make a globe for a social studies project. Rounded to the nearest foot, how much greater is the surface area of a globe with a radius of 4 feet than one with a radius of 2 feet?

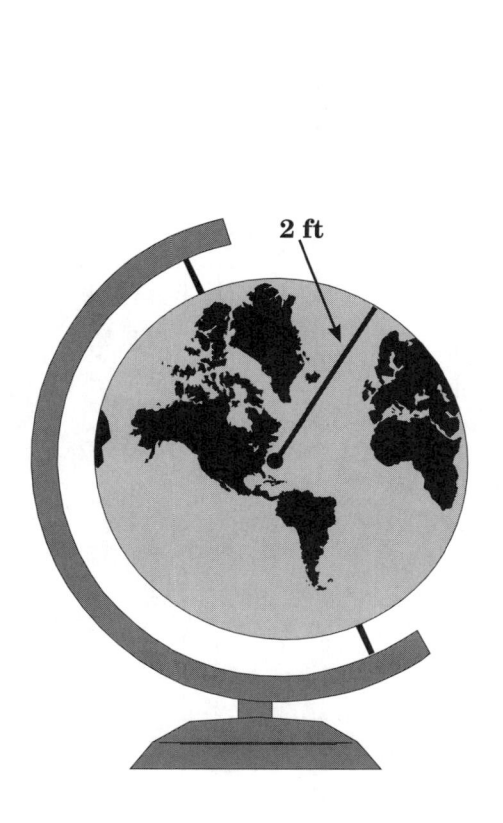

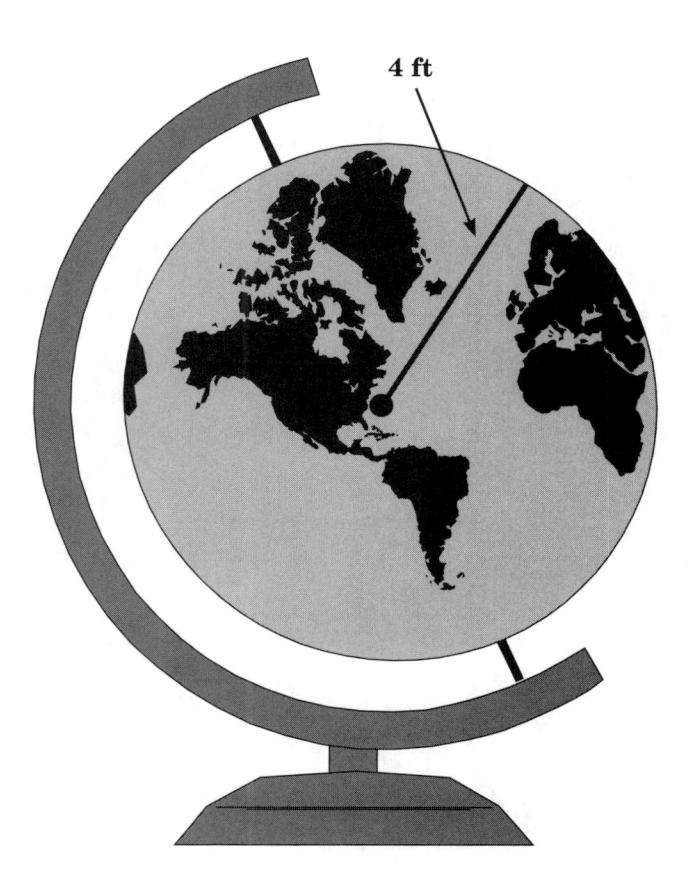

F 25 ft²

G 96 ft²

H 144 ft²

J 151 ft²

45. Last week, the sales each day at a deli were $250, $530, $300, $424, $506, $290, and $300. Select the most effective measure to convince a potential investor that the deli has high sales.

A Mean

B Median

C Mode

D Range

GO ON

46. Which graph best represents the inequality $y \geq -\frac{1}{2}x + 2$?

F

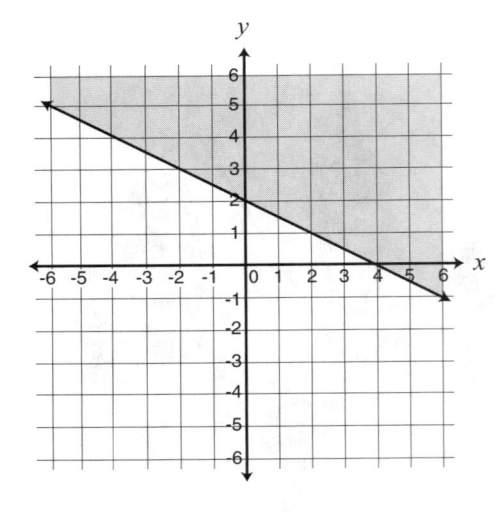

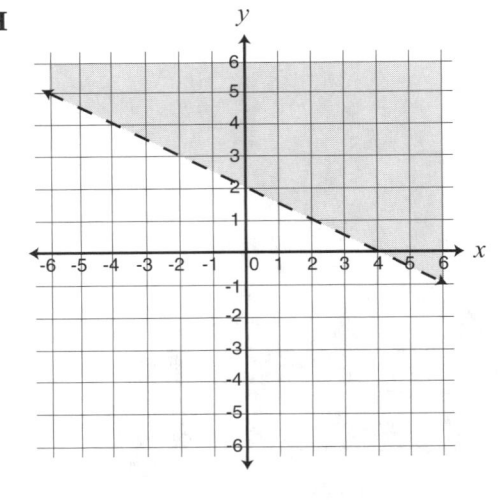

G

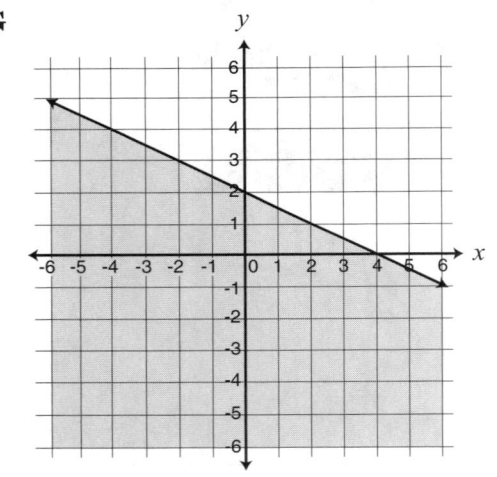

H

J

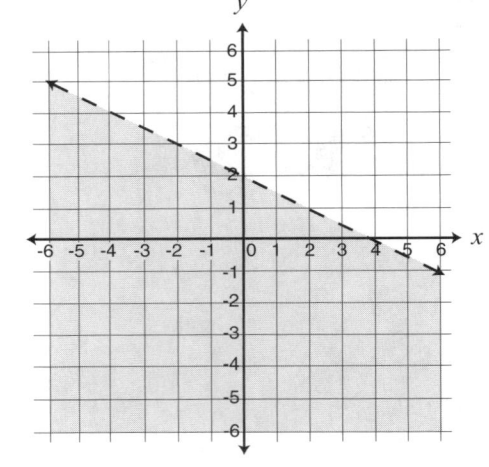

47. Tamara conducted an experiment and recorded the data in the table shown below.

x	y
1	-1
2	2
3	7
4	14

Which equation best describes these data?

A $y = x^2 - 2$

B $y = x^2$

C $y = x^2 - 1$

D $y = x - 3$

48. To entertain at a birthday party, a magician charges $200 plus $5 per guest. Which best represents c, the total cost to have the magician entertain at a birthday party, with g being the number of guests?

F $c = 200g + 5$

G $c = \dfrac{200}{5g}$

H $c = 200 + 5 + g$

J $c = 200 + 5g$

49. Find the x- and y-intercepts of $5x - \dfrac{1}{4}y = 12$.

A x-intercept (60, 0)
   y-intercept (0, −3)

B x-intercept (0, −3)
   y-intercept (60, 0)

C x-intercept $(\dfrac{12}{5}, 0)$
   y-intercept (0, −48)

D x-intercept (0, −48)
   y-intercept $(\dfrac{12}{5}, 0)$

50. Which equation will produce the widest parabola when graphed?

F $y = -4x^2 + 5$

G $y = -\dfrac{1}{2}x^2 + 5$

H $y = 2x^2 + 5$

J $y = 4x^2 + 5$

51. What is the seventh number in this sequence?

1, −2, 4, −8, 16,...

A −128

B −64

C 64

D 128

52. What is the probability of flipping 4 coins and getting all heads or all tails?

   F $\dfrac{1}{16}$

   G $\dfrac{1}{8}$

   H $\dfrac{1}{4}$

   J $\dfrac{1}{2}$

53. Which equation corresponds to the line graphed below?

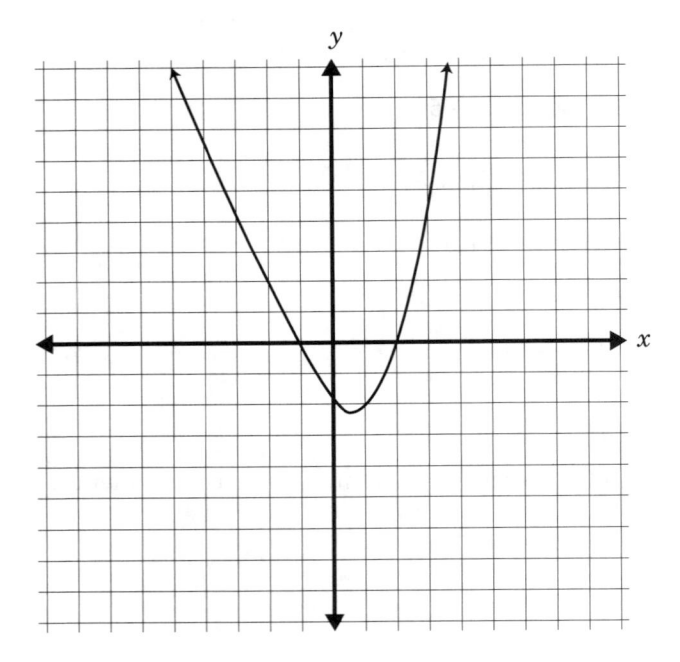

   A $y = x^2 - x + 2$

   B $y = x^2 - x - 2$

   C $y = x^2 - 3x - 2$

   D $y = x^2 - 2$

54. Pilar spent a total of $200 for 6 sweaters. Later she bought another sweater. She spent an average of $32.00 per sweater for the seven sweaters. What did she pay for the seventh sweater?

   F $24.00

   G $28.00

   H $32.00

   J $36.00

55. The number of students enrolled in introductory courses at a university is shown on the graph below.

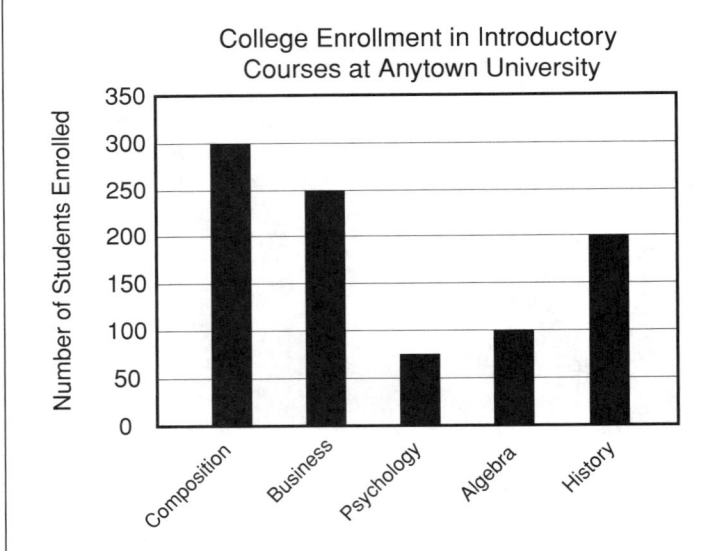

   How many more students are enrolled in Composition than in Psychology?

   A 200

   B 225

   C 250

   D 375

56. If $a + b = c$ and $d = e$, which of the following must be true?

   F $a + b + d = c + e$
   G $a + b - d = c - e - d$
   H $a + c = b + d + e$
   J $a + c - b = d + e$

57. If y varies directly with x and y is 21 when x is 7, which of the following represents the situation?

   A $y = x + 7$
   B $y = 3x$
   C $y = x^2$
   D $y = \dfrac{5}{2}x$

58. Which of the following 3-dimensional figures has one vertex and a base in which all points are equidistant from one point?

   F Sphere
   G Cylinder
   H Cone
   J Cube

59. What is the slope of the line graphed below?

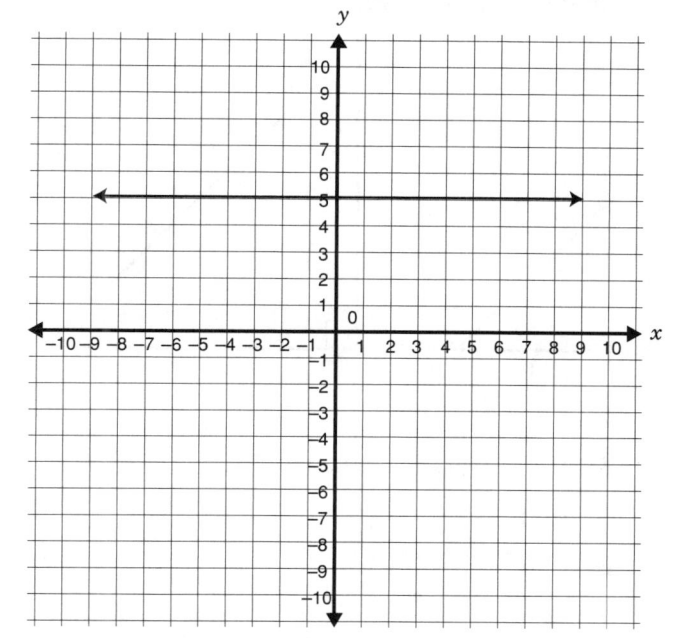

   A -1
   B 0
   C $\dfrac{1}{2}$
   D Undefined

GO ON

60. Find the missing length (x) of the pair of similar figures shown below.

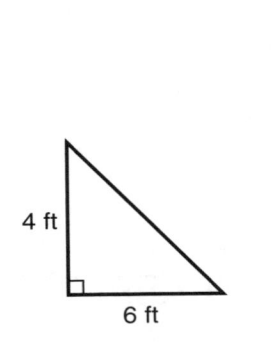

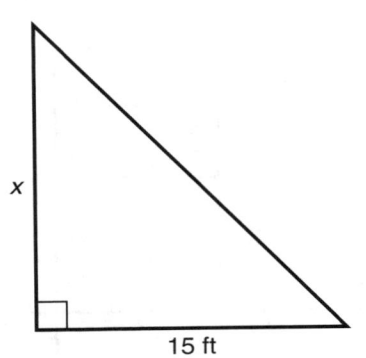

F 8 ft
G 10 ft
H 12 ft
J 18 ft

BE SURE YOU HAVE RECORDED ALL OF YOUR ANSWERS ON THE ANSWER DOCUMENT

Practice Test 1

1. Ⓐ Ⓑ Ⓒ Ⓓ
2. Ⓕ Ⓖ Ⓗ Ⓙ
3. Ⓐ Ⓑ Ⓒ Ⓓ
4. Ⓕ Ⓖ Ⓗ Ⓙ
5. Ⓐ Ⓑ Ⓒ Ⓓ
6. Ⓕ Ⓖ Ⓗ Ⓙ
7. Ⓐ Ⓑ Ⓒ Ⓓ
8. Ⓕ Ⓖ Ⓗ Ⓙ
9. Ⓐ Ⓑ Ⓒ Ⓓ
10. Ⓕ Ⓖ Ⓗ Ⓙ
11. Ⓐ Ⓑ Ⓒ Ⓓ
12.

13. Ⓐ Ⓑ Ⓒ Ⓓ
14. Ⓕ Ⓖ Ⓗ Ⓙ
15. Ⓐ Ⓑ Ⓒ Ⓓ
16. Ⓕ Ⓖ Ⓗ Ⓙ
17. Ⓐ Ⓑ Ⓒ Ⓓ
18. Ⓕ Ⓖ Ⓗ Ⓙ
19. Ⓐ Ⓑ Ⓒ Ⓓ

20. Ⓕ Ⓖ Ⓗ Ⓙ
21. Ⓐ Ⓑ Ⓒ Ⓓ
22. Ⓕ Ⓖ Ⓗ Ⓙ
23. Ⓐ Ⓑ Ⓒ Ⓓ
24. Ⓕ Ⓖ Ⓗ Ⓙ
25.

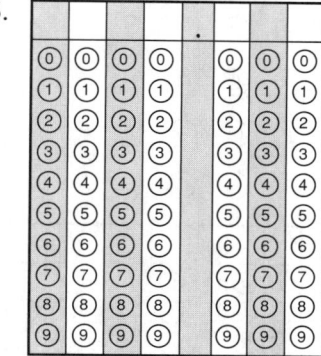

26. Ⓕ Ⓖ Ⓗ Ⓙ
27. Ⓐ Ⓑ Ⓒ Ⓓ
28. Ⓕ Ⓖ Ⓗ Ⓙ
29.

30. Ⓕ Ⓖ Ⓗ Ⓙ

TAKS — Exit Level Mathematics

31. Ⓐ Ⓑ Ⓒ Ⓓ
32. Ⓕ Ⓖ Ⓗ Ⓙ
33. Ⓐ Ⓑ Ⓒ Ⓓ
34. Ⓕ Ⓖ Ⓗ Ⓙ
35. Ⓐ Ⓑ Ⓒ Ⓓ
36. Ⓕ Ⓖ Ⓗ Ⓙ
37. Ⓐ Ⓑ Ⓒ Ⓓ
38. Ⓕ Ⓖ Ⓗ Ⓙ
39. Ⓐ Ⓑ Ⓒ Ⓓ
40. Ⓕ Ⓖ Ⓗ Ⓙ
41. Ⓐ Ⓑ Ⓒ Ⓓ
42. [grid-in answer box]

43. Ⓐ Ⓑ Ⓒ Ⓓ
44. Ⓕ Ⓖ Ⓗ Ⓙ
45. Ⓐ Ⓑ Ⓒ Ⓓ
46. Ⓕ Ⓖ Ⓗ Ⓙ
47. Ⓐ Ⓑ Ⓒ Ⓓ
48. Ⓕ Ⓖ Ⓗ Ⓙ
49. Ⓐ Ⓑ Ⓒ Ⓓ
50. Ⓕ Ⓖ Ⓗ Ⓙ
51. Ⓐ Ⓑ Ⓒ Ⓓ
52. Ⓕ Ⓖ Ⓗ Ⓙ
53. Ⓐ Ⓑ Ⓒ Ⓓ
54. Ⓕ Ⓖ Ⓗ Ⓙ
55. Ⓐ Ⓑ Ⓒ Ⓓ
56. Ⓕ Ⓖ Ⓗ Ⓙ
57. Ⓐ Ⓑ Ⓒ Ⓓ
58. Ⓕ Ⓖ Ⓗ Ⓙ
59. Ⓐ Ⓑ Ⓒ Ⓓ
60. Ⓕ Ⓖ Ⓗ Ⓙ

BE SURE YOU HAVE RECORDED ALL OF YOUR ANSWERS ON THE ANSWER DOCUMENT

STOP

Practice Test 1 Answer Explanations

1. A   A(b)(1)(A)
The number of people with cell phones varies based on the year, so the year is the independent quantity.

2. G   A(d)(2A)
If the amount of pollution was 512,000 tons in 2005, multiply by $\frac{1}{2}$ for each year until 2012: $512,000 \times (\frac{1}{2})^7 = 4,000$ tons.

3. A   08   G(f)(1)(A)
Remember that a reflection is a mirror image. Therefore, when the triangle is reflected over the x-axis, the y-coordinates will change sign (the x-coordinates will stay the same), and it will look as if it's upside-down.

4. F   02   A(b)(4)(A)
If you substitute the values in parentheses for x, the independent variable, you'll see that the only value that is corresponding is -13. Specifically, when $x = -4$, $y = -13$.

5. B   09   (8.3)(B)
Since the ratio of true answers to false answers is 4:3, the ratio of true answers to all answers is 4:7. Convert 4:7 to a percent by multiplying $\frac{4}{7}$ by 100 and write the percent sign. $\left(\frac{4}{7}\right)\left(\frac{100}{1}\right)\% = 57\%$, when rounded off to the nearest integer.

6. H   (8.13)(B)
Fish Fingers received fewer votes than the other options, so it is the least favorite meal. The other possible choices are not valid conclusions.

7. B   03   A(c)(2)(A)
Use any two pairs of coordinates on the line to find the slope, by using the point-slope formula of an equation from your Mathematics Chart. Specifically, you can use the points (2, 2) and (4, 1). The slope equals $(1-2)/(4-2) = -\frac{1}{2}$.

8. G   (8.11)(B)
To find the probability, you need to determine the proportion of orange tiles to the total number of tiles. Add up all the tiles, and use this number (40) as the denominator. Then simplify: $\frac{5}{40} = \frac{1}{8}$.

9. C   A(d)(2)(B)
To solve this question, replace the x value with the value in the ordered pair, and find which equation makes $f(x) = 0$. For answer choice A, the equation equals 30. For answer choice B, the equation equals 20. For answer choice C, the equation equals 0. For answer choice D, the equation equals 12. Therefore, answer choice C is correct.

10. J   A(c)(4)(B)
Sketch the new equation ($y = \frac{1}{3}x + 1$) on the graph first. The y-intercept is +1 so make a point at the place (0, 1) on the graph. The slope is $\frac{1}{3}$ so the line will rise (y) one unit and run (x) three units, so another point can be at (–3, 0). Draw a straight line to connect the dots and extend it across the entire graph. The lines intersect at (3, 2). That is the solution to the set of equations.

11. A   01   A(b)(1)(C)
If the candle manufacturer charges $12.00 plus 3% or .03 shipping and handling, the correct equation is $c = \$12.00y + .03s$.

12. 447 sq. cm.   08   G(e)(1)(C)
The area of the rectangular cross section is found by multiplying the length and width. To find the length, use the Pythagorean theorem. Here, $= c$.

$10^2 + 20^2 = 500$

$\sqrt{500} \approx 22.36$

$22.36 \times 20 \approx 447$ sq. cm

13. C   07   G(d)(2)(B)
Answer choice C shows a curve that is symmetrical with the x-axis only.

Practice Test 1 Answer Explanations

14. J   01   A(b)(1)(D)

The interest on Sharon's investment of $4,000 at an annual rate of 3% could be represented like this: 4000 × 0.03 or 0.03(4000). The interest on the money she invested at a rate of 5.50% could be represented by using the variable x, since we don't know how much money she invested at this rate: 0.055x. So the total interest, y, is the sum $y = 0.03 (4000) + 0.055x$.

15. A   07   G(d)(1)(B)

An octahedron has eight triangular faces, six vertices, and twelve edges.

16. G   06   G(e)(3)(A)

If you look carefully at parallelogram DEFG, you'll see that its coordinates are (1, 3), (2, 7), (6, 3), and (7, 7). The coordinates of KLMN need to be a translation 10 units down the y-axis and 8 units to the left along the x-axis. Point N corresponds to point D, which is (2, 7), so the coordinates of point N will be (2 – 8, 7 – 10), or (–6, –3).

17. A   04   A(c)(4)(A)

To answer this question, you need to carefully consider the information you are given. You know that there are 85 gold and silver memberships, so one of the equations should be $g + s = 85$. The second equation should contain the value of each membership: $640g + 400s = 140,000$.

18. J   09   (8.12)(C)

To answer this question, you need to determine the percentage of votes for each band. To do this, add up all of the votes, and use this number (550) as the denominator. Then divide the denominator into the numerators and multiply by 100% to find the percentage. For example, for Band 1, $\frac{280}{550} \approx .51$; and .51 × 100% is 51%.

19. A   02   A(b)(4)(B)

To simplify the expression, multiply out each term to get $6x - 3y + 8x + 4y + 3x + 3y$, and then combine like terms to get $17x + 4y$.

20. F   08   G(e)(1)(A)

To find the area of the polygon, you need to multiply the length by the height; in this case, the answer is n^2.

21. D   03   A(c)(1)(B)
When graphed, a linear function is a non-vertical straight line. Answer choice D is correct.

22. F   05   A(d)(3)(A)
Write out each number and add. When you add 1,180,000 and 1,950,000, the answer is 3,130,000. To express the answer in scientific notation, you would move the decimal six places to the left, so the exponent of 10 is still 6.

23. D   06   G(c)(1)(B)
To answer this question, you need to remember that a full rotation is 360 degrees, so each of the eight points on the dial represents one-eighth of 360 degrees, or 45 degrees. So the lock is moved 90 degrees counterclockwise and then back 225 degrees, and C to F is the only answer choice that reflects the resulting 135-degree clockwise rotation.

24. G   G(c)(1)(C)
A right isosceles triangle has one right angle and two angles with equal measures. To find the measure of the two equal angles, subtract 90° from the total number of degrees in a right triangle (180°), then divide by two: (180° − 90°) ÷ 2 = 45°.

25. $405.30   10   (8.14)(A)
Shamus's hours total 45.5. His pay for 40 hours at $10/hour is $400.00. His hourly overtime rate is $15.00 an hour. He worked 5.5 hours of overtime, so he earned an additional 5.5 × $15 = $82.50 in overtime pay. His weekly salary plus his overtime is therefore $482.50. The amount of money withheld for taxes is $482.50 × .16 = $77.20. So Shamus takes home $482.50 − $77.20 = $405.30.

26. J   06   G(c)(1)(A)
In an isosceles trapezoid, the base angles are congruent.

27. C   A(b)(2)(D)
The dots do not approximate a line, so there is no trend between the amount of coffee consumed and test scores.

28. F   04   A(c)(3)(A)
To solve this problem, remember that the salesperson earns $325, a flat fee, in addition to $\frac{1}{6}$ of her total sales. Therefore, the correct answer choice is F: $y = 325 + \frac{1}{6}x$.

29. 35°C   04   A(c)(3)(B)

Substitute F = 95 into the equation to get $95 = \frac{9}{5}C + 32$. Subtract 32 from each side to get $63 = \frac{9}{5}C$, and then multiply each side by 5 : 315 = 9C, or C = 35°.

30. F   06   G(b)(4)(A)

A reflection is a mirror image. Only answer choice F shows a pattern made entirely from reflections.

31. C   10   (8.16)(B)

Since the mean is the total divided by the number of days, the total = mean × number of days, so to find out how many people attended the museum over five days, multiply the mean by 5.

32. H   08   G(e)(1)(A)

To find the area of the figure, you need to multiply the length times the width for the two rectangles that compose the figure. The left rectangle has height *a* and width *f*, or an area *af*. The right rectangle has height *c* and width *d*, or an area *cd*. So the total area is *af* + *cd*.

33. C   03   A(c)(2)(D)

If you substitute values from two coordinates on the line into each of these equations, you'll see that answer choice C is the only correct one. For example, you could substitute the points (2, 2) and (3, 4).

34. G   07   G(d)(1)(B)

The net in answer choice G is the only one that has six faces.

35. C   09   (8.16)(A)

The graph increases by 50 each year. So there will be 900 college graduates in Amy's town in 2008.

36. F   07   G(d)(2)(B)

When parallel lines are intersected by a transversal, opposite inside angles are congruent.

37. A A(b)(3)(A)
The only expression that is the same in this situation is A, since Jordan's age, j, is equal to Cayce's age, c, multiplied by three and then minus two. The other answer choices show the wrong sequence of operations, or operations being applied to the wrong variable.

38. H 06 G(c)(1)(A)
Angle CAB and angle BAD measure 90 degrees together. Therefore, you can find the measure of angle CAB if you subtract 35 from 90.

39. B 04 A(c)(3)(A)
The sign in the inequality should be equal to or less than, since Kaitlyn can't spend more than $80, but she can spend exactly $80.

40. F 05 A(d)(1)(C)
If the graph is translated 7 units down, you would subtract 7 from each y value, so the equation would be $y = x^2 - 7$.

41. C 02 A(b)(2)(A)
The graph is a parabola, so the parent function is a quadratic.

42. 60 02 A(b)(2)(C)
According to the pattern, there are 12 dancers in the first circle, 14 in the second, 16 in the third, and 18 in the fourth. The total number of dancers is 60.

43. C 08 G(e)(1)(D)
If you multiply 12 by 6 and then the answer by 4, you get 288.

44. J G(f)(1)(D)
To find the surface area of each globe, substitute the radius into the formula for the surface area of a sphere. For the smaller globe, the surface area is $4\pi(2^2) = 50$ ft². For the larger globe, the surface area is $4\pi(4^2) = 201$ ft². Then subtract to find the difference: 201 ft² − 50 ft² = 151 ft².

45. A 09 (8.12)(A)
The mean sales are about $370. Both the mode and median are $300, and the range is $280. Therefore, the mean gives the highest sales.

Practice Test 1 Answer Explanations   251

46. F   01   A(b)(1)(D)
Answer choice F shows the correct graph with a solid line because the sign ≥ means it also includes the points on the line. The region shaded is above the line.

47. A   01   A(b)(1)(B)
Substitute the given values into each equation to choose the correct one. If you plug the values 1, 2, 3, 4 for x in the equation in answer choice A, you'll see that this answer choice is correct.

48. J   04   A(c)(3)(A)
To solve this problem, you need to choose the equation in which $5, the cost per guest, is multiplied by g, the unknown number of guests, and this is added to the fixed fee of $200.

49. C   A(c)(2)(E)
To find the x-intercept, substitute $y = 0$ and solve: $5x - \frac{1}{4}(0) = 12 = 5x; x = \frac{12}{5}$. The x-intercept is $(\frac{12}{5}, 0)$. To find the y-intercept, substitute $x = 0$ and solve: $5(0) - \frac{1}{4}y = 12 = -\frac{1}{4}y; y = -48$. The y-intercept is $(0, -48)$.

50. G   A(d)(1)(B)
The width of a parabola is determined by the coefficient of the x^2 variable. The lowest absolute value of this coefficient will have the widest arc, and the highest absolute value will have the narrowest arc. Choice G has the lowest absolute value for this coefficient, so it will have the widest arc.

51. C   (8.14)(C)
To find the value of a number in this sequence, multiply the number before it by –2. Multiply the fifth number by –2 twice to find the seventh number: $16 \times -2 \times -2 = 64$.

52. G   09   (8.11)(A)
There is one instance of all tails (TTTT) and one instance of all heads (HHHH) out of 16 possibilities. The question asks for the probability of either, so the correct answer is $\frac{2}{16} = \frac{1}{8}$.

53. B A(d)(1)(D)
Substitute points from the line into each equation to determine whether the values are solutions to the equation. The point (2, 0) is on the line and satisfies answer choice B: $0 = 2^2 - 2 - 2$. The point (-1, 0) from the line also satisfies answer choice B: $0 = (-1)^2 - (-1) - 2$. The point (2, 0) also satisfies answer choice A, but (-1, 0) does not. Neither point satisfies answer choices C or D. Therefore, B is the answer.

54. F 09 (8.3)(B)
To find the answer to this problem, multiply the average cost for the seven sweaters, $32.00, by 7. This will give the total cost for all seven sweaters. Then deduct $200 from this amount to get the cost of the seventh sweater.

55. B 09 (8.12)
Three hundred students are enrolled in composition and 75 are enrolled in psychology. If you subtract 75 from 300, the answer is 225.

56. F 10 (8.16)(B)
Answer choice F is true according to the addition property of equality.

57. B 03 A(c)(2)(G)
If x is 7 and y is 21, x is multiplied by 3 to get y, so the correct answer is $y = 3x$.

58. H 07 G(e)(2)(D)
The description of the 3-dimensional figure is a cone. It is the only one of the answer choices that has one vertex and a circular base.

59. B 03 A(c)(2)(A)
A line parallel to the x-axis has a slope of zero. If you didn't remember this, you could substitute two sets of coordinates, such as (-4, 5) and (5, 5) into the formula $\text{Slope} = \frac{y_2 - y_1}{x_2 - x_1}$, which would give 0 for all points.

60. G 08 G(f)(1)(B)
To find x, set up a proportion $\left(\frac{4}{6} = \frac{x}{15}\right)$ and cross-multiply to solve for x.

TAKS Mathematics Practice Test

2

Directions: This Practice Test contains 60 questions. A graphing calculator may be used.

Mark answers in the Answer Document at the end of this test.

Mathematics Chart – Measurements

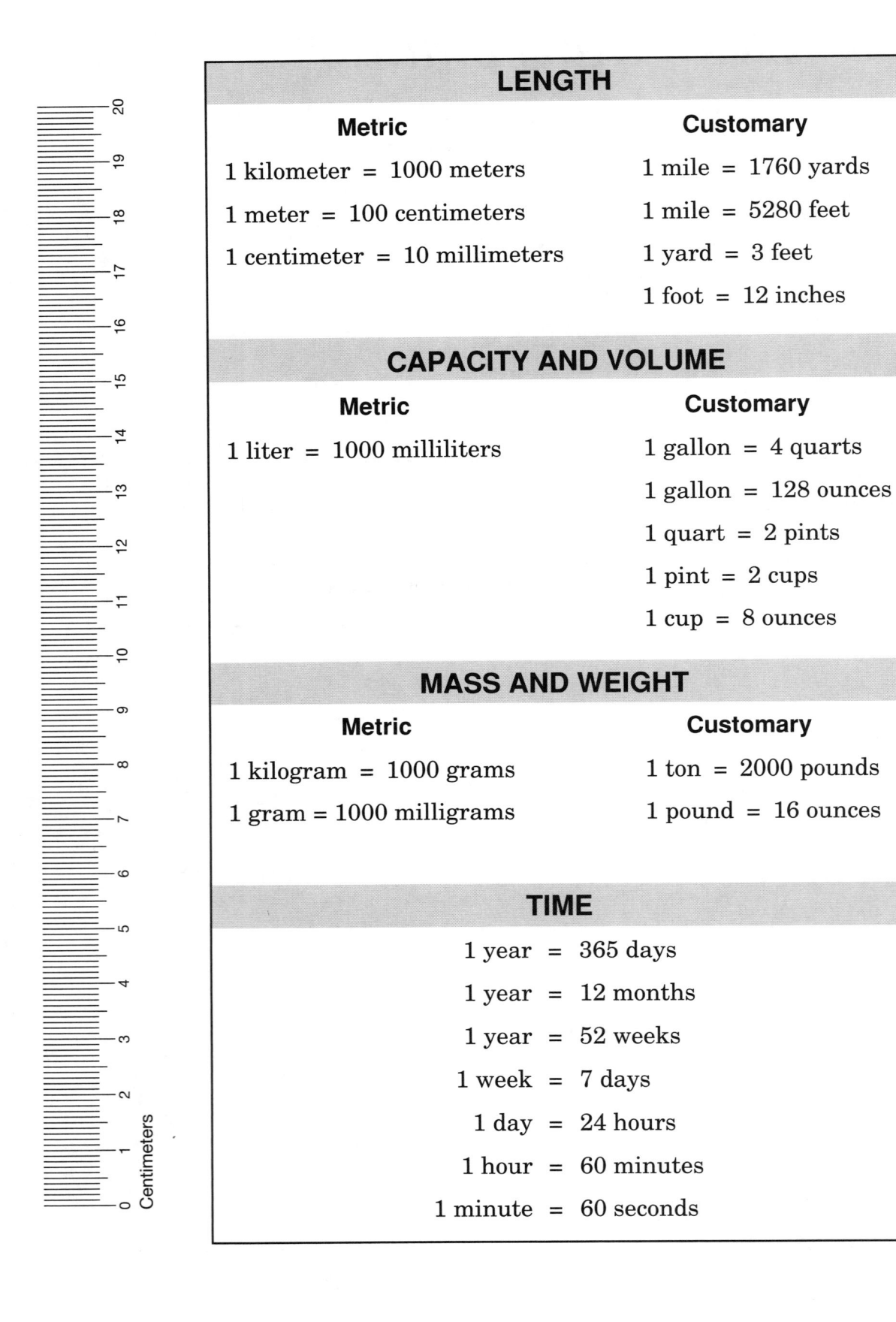

LENGTH

Metric

1 kilometer = 1000 meters

1 meter = 100 centimeters

1 centimeter = 10 millimeters

Customary

1 mile = 1760 yards

1 mile = 5280 feet

1 yard = 3 feet

1 foot = 12 inches

CAPACITY AND VOLUME

Metric

1 liter = 1000 milliliters

Customary

1 gallon = 4 quarts

1 gallon = 128 ounces

1 quart = 2 pints

1 pint = 2 cups

1 cup = 8 ounces

MASS AND WEIGHT

Metric

1 kilogram = 1000 grams

1 gram = 1000 milligrams

Customary

1 ton = 2000 pounds

1 pound = 16 ounces

TIME

1 year = 365 days

1 year = 12 months

1 year = 52 weeks

1 week = 7 days

1 day = 24 hours

1 hour = 60 minutes

1 minute = 60 seconds

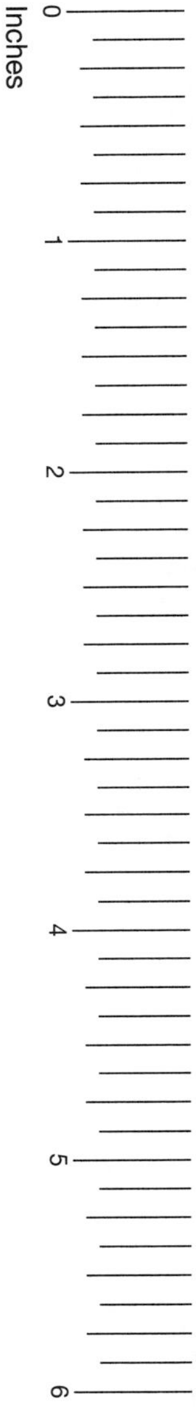

Mathematics Chart – Formulas

Perimeter	rectangle	$P = 2l + 2w$ or $P = 2(l + w)$
Circumference	circle	$C = 2\pi r$
Area	rectangle	$A = lw$ or $A = bh$
	triangle	$A = \dfrac{1}{2}bh$ or $A = \dfrac{bh}{2}$
	trapezoid	$A = \dfrac{1}{2}(b_1 + b_2)h$ or $\dfrac{(b_1 + b_2)h}{2}$
	circle	$A = \pi r^2$
Surface Area	cube	$S = 6s^2$
	cylinder (lateral)	$S = 2\pi rh$
	cylinder (total)	$S = 2\pi rh + 2\pi r^2$ or $S = 2\pi r(h + r)$
	cone (lateral)	$S = \pi rl$
	cone (total)	$S = \pi rl + \pi r^2$ or $S = \pi r(l + r)$
	sphere	$S = 4\pi r^2$
Volume	prism or cylinder	$V = Bh^*$
	pyramid or cone	$V = \dfrac{1}{3}Bh^*$
	sphere	$V = \dfrac{4}{3}\pi r^3$

*B represents the area of the base of a solid figure.

Pi	π	$\pi \approx 3.14$ or $\pi \approx \dfrac{22}{7}$
Pythagorean Theorem		$a^2 + b^2 = c^2$
Distance Formula		$d = \sqrt{(x_2 - x_1)^2 + (y_2 - y_1)^2}$
Slope of a Line		$m = \dfrac{y_2 - y_1}{x_2 - x_1}$
Midpoint Formula		$M = \left(\dfrac{x_1 + x_2}{2}, \dfrac{y_1 + y_2}{2}\right)$
Quadratic Formula		$x = \dfrac{-b \pm \sqrt{b^2 - 4ac}}{2a}$
Slope-Intercept Form of an Equation		$y = mx + b$
Point-Slope Form of an Equation		$y - y_1 = m(x - x_1)$
Standard Form of an Equation		$Ax + By = C$
Simple Interest Formula		$I = prt$

1. If $\overline{XY}$ = 10 feet and $\overline{XZ}$ = 12 feet, what is the area of triangle XYZ?

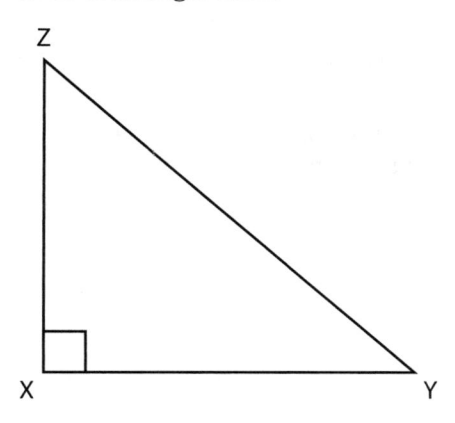

   A 40 ft²

   B 60 ft²

   C 120 ft²

   D 140 ft²

2. Frankie had 30 baseball cards at the beginning of the week. If x represents the number of basketball cards Frankie gave to his friend Amy on Tuesday, and y represents the number of baseball cards his mother gave him on Thursday, which expression shows the number of baseball cards Frankie had at the end of the week?

   F $x + 30 - y$

   G $30 - x + y$

   H $30x - y$

   J $30y + x$

3. A rotating sprinkler is used to water a yard. The radius of the area being sprayed is 12 feet. What is the approximate wet area of the yard?

   A 38 ft²

   B 120 ft²

   C 144 ft²

   D 452 ft²

4. A sphere has a surface area of 201m². What is the approximate radius of the sphere?

   Record your answer and fill in the bubbles on your answer document. Be sure to use the correct place value.

5. A contractor uses a 15-foot ladder to reach the roof. The ladder is 9 feet away from the house and forms a right triangle.

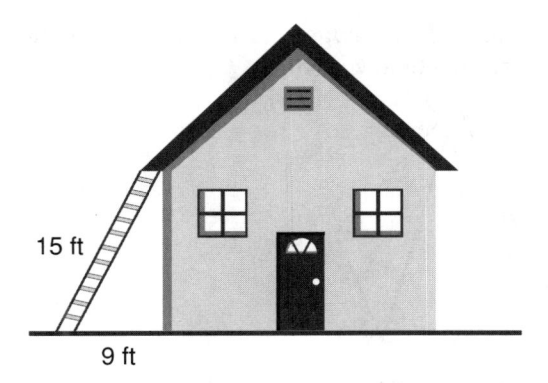

   How high up the house is the ladder?

   Record your answer and fill in the bubbles on your answer document. Be sure to use the correct place value.

6. What is the volume of a cube that has an edge of 4 inches?

   F 12 in³
   G 16 in³
   H 64 in³
   J 96 in³

7. A biologist noticed that the population of butterflies in a sample doubled every 5 days. If the initial population sample was 100 butterflies, what was the population of butterflies at the end of 10 days?

   A 200
   B 300
   C 400
   D 500

8. How many faces, edges, and vertices does the solid shown below have?

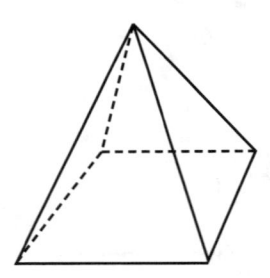

   F 5 faces, 8 edges, and 5 vertices
   G 6 faces, 8 edges, and 6 vertices
   H 5 faces, 9 edges, and 5 vertices
   J 4 faces, 4 edges, and 6 vertices

9. Jeffrey needs to simplify the following expression for his homework assignment.

   $$4(x + 2y) + 2(3x - y) - (x + y)$$

   Which of the following expressions is equivalent to the expression above?

   A $10x - 7y$
   B $9x + 5y$
   C $-12x$
   D $-12x + 7y$

10. Ray works for his older brother Cory at a drugstore in their neighborhood. For every dollar Ray earns, Cory earns 1.5 times this amount. This week, the two earned $750. How much did Cory earn?

    Record your answer and fill in the bubbles on your answer document. Be sure to use the correct place value.

11. The graph below represents which type of parent function?

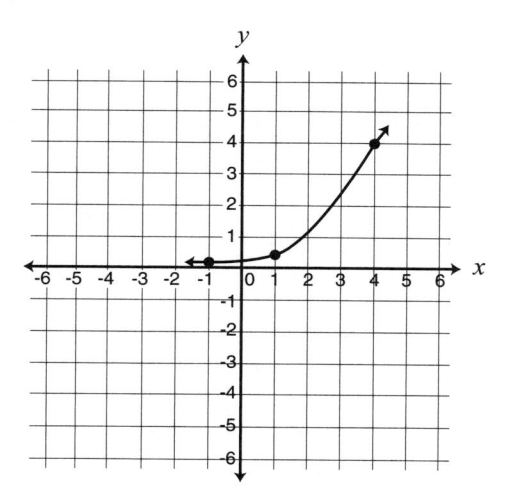

A Exponential
B Absolute value
C Linear
D Quadratic

12. Peter bought a mountain bike for $195.99. When he took the bike home, he noticed a large scratch on one side. The bike shop gave Peter a 35% refund on the bike. How much did Peter pay for the bike after the refund?

Record your answer and fill in the bubbles on your answer document. Be sure to use the correct place value.

13. What is another way to express 225?

A 5^4
B 15^2
C 22.5×10^2
D 12^3

14. Simplify the expression below.

$$\frac{3y}{y^3}$$

F $3y^2$

G $\frac{3y}{y}$

H $\frac{3}{y^2}$

J $\frac{3}{y^3}$

15. Rubina purchased 2 pairs of sneakers for a total of $84.99. The sneakers were discounted 20% the next week, and the store manager agreed to give Rubina a refund equal to the amount of the discount. How much did the sneakers finally cost Rubina? Show your work in the space below.

Record your answer and fill in the bubbles on your answer document. Be sure to use the correct place value.

16. In 2006, 136 million people in the United States were employed. Of these, about 13% worked in manufacturing jobs. According to this information, about how many millions of people in the United States were employed in manufacturing jobs?

Record your answer and fill in the bubbles on your answer document. Be sure to use the correct place value.

GO ON

17. Melanie uses the expression $8a + 12b$ to determine the amount she earns at a pay rate of $8 an hour for regular hours, a, plus time and a half for overtime hours, b. One week she worked 40 hours, plus 8 hours of overtime. What is her total pay for the week?

Record your answer and fill in the bubbles on your answer document. Be sure to use the correct place value.

18. Which graph best represents the inequality $y < 3x - 1$?

F

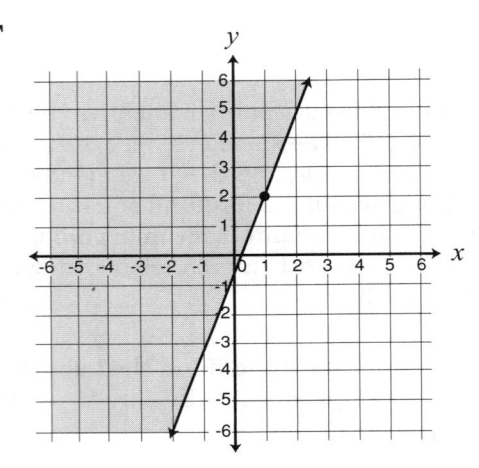

H

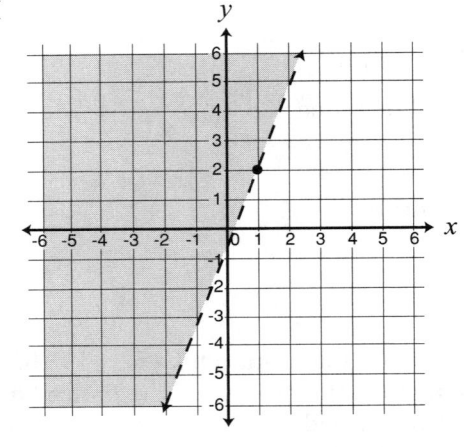

G

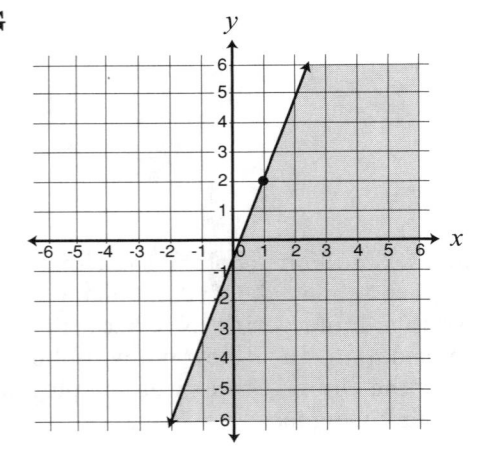

J

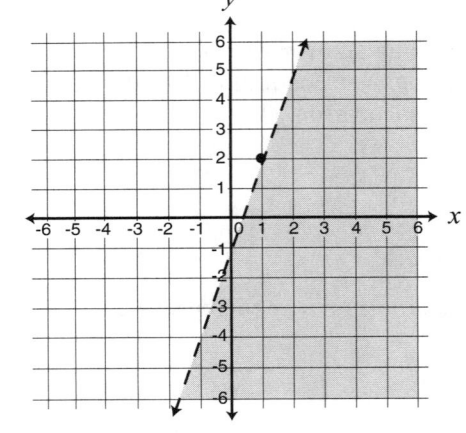

19. Given the inequality $6y < 42$, solve for y.

   A $y = 7$

   B $y < 7$

   C $y > 7$

   D $y \leq 7$

20. If a commercial jet travels 570 miles per hour, about how many miles will it travel in 4 hours?

   F 143 miles

   G 1,140 miles

   H 2,280 miles

   J 2,850 miles

21. If a jacket originally cost $75 and is selling at a 20% discount, what is the amount of the discount?

   A $7.50

   B $11.25

   C $15.00

   D $18.75

22. If $\angle ACB$ measures 45°, what is the measure of $\angle BCD$?

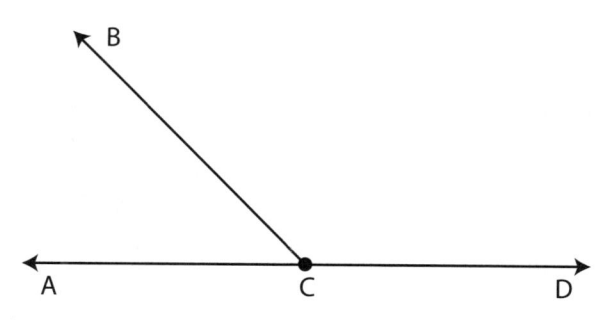

   F 40°

   G 45°

   H 90°

   J 135°

23. Madeline earns $8 an hour for babysitting her cousins during the 10 weeks of summer vacation. If she averages 12 hours per week, what is a reasonable estimate of what Madeline earns during the summer?

   A $80

   B $96

   C $120

   D $960

24. The oldest person in an audience of 100 people is 62. If the range is 51, what is the age of the youngest member of the audience?

   F 10

   G 11

   H 12

   J 13

GO ON

25. When an object that weighs y pounds is hung from a spring, the spring stretches x inches, as shown in the picture below.

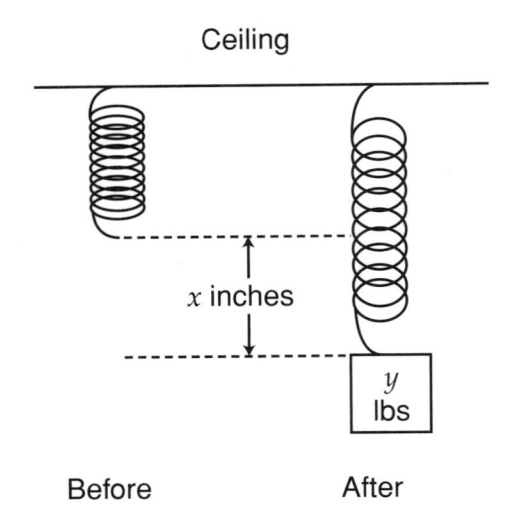

Ceiling

x inches

y lbs

Before   After

Use the equation below to determine how many inches the spring will stretch if an object weighing 10 pounds is attached to the end of the string.

$$y = \left(\frac{1}{4}\right)x$$

Record your answer and fill in the bubbles on your answer document. Be sure to use the correct place value.

26. In the figure, the letter "R" is to be first reflected over the vertical line a and then slid across the horizontal line b.

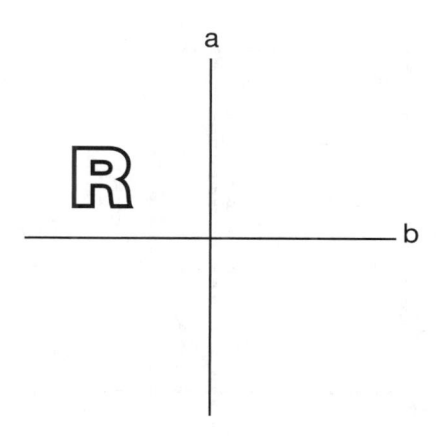

Which of these figures would be the correct orientation of the letter "R" after the two transformations described?

F

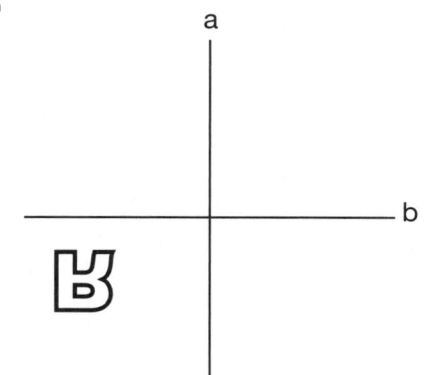

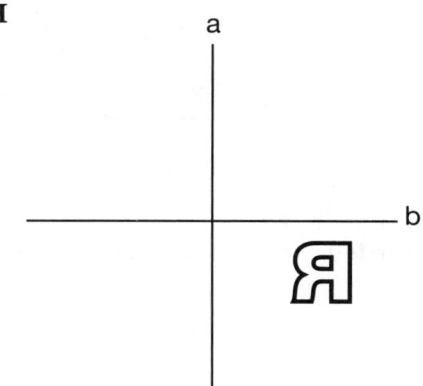

G

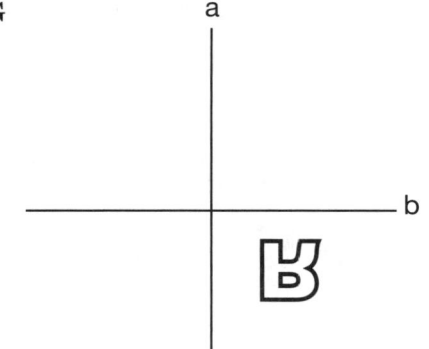

H

J

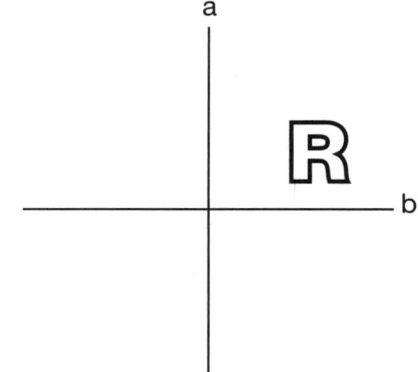

27. The graph of line d is shown below. The slope of this line is $\frac{3}{2}$.

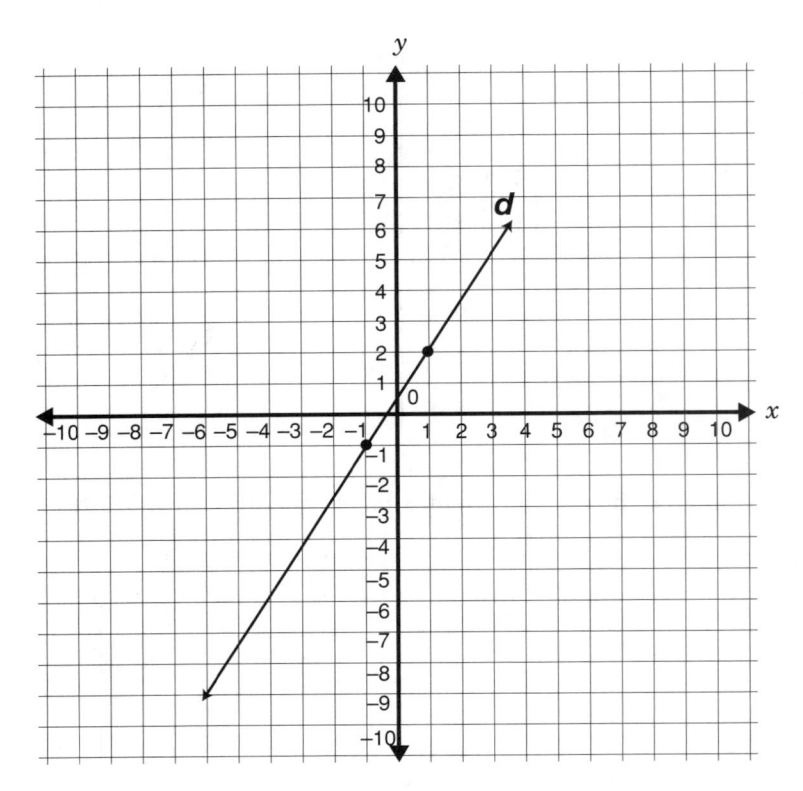

Which of the following is the slope of a line that is parallel to line d?

A $-\frac{3}{2}$

B 0

C $\frac{1}{2}$

D $\frac{3}{2}$

28. Which graph represents a linear function?

F

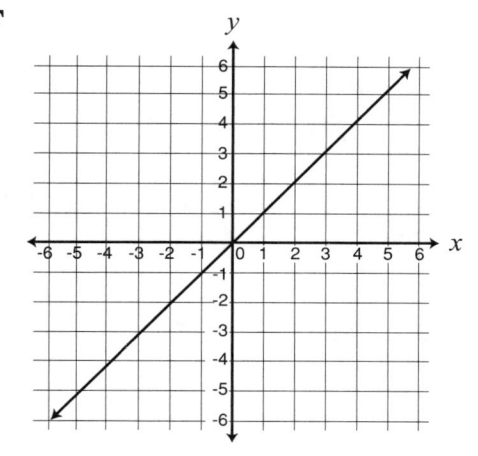

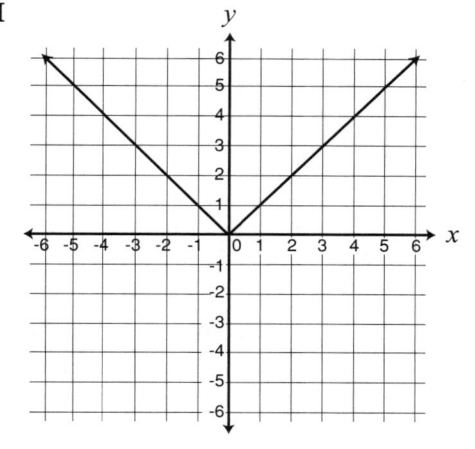

G

H

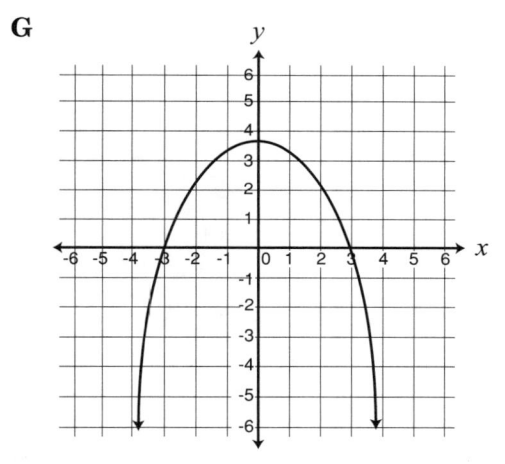

J

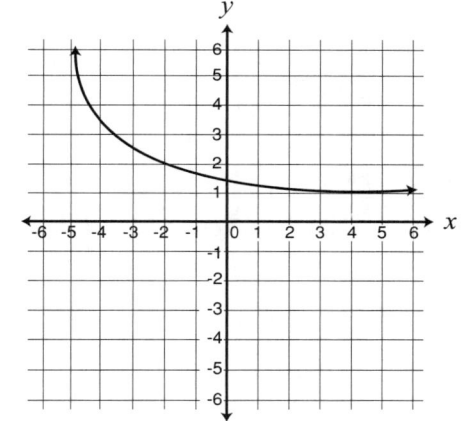

29. What is the slope of the line in the graph below?

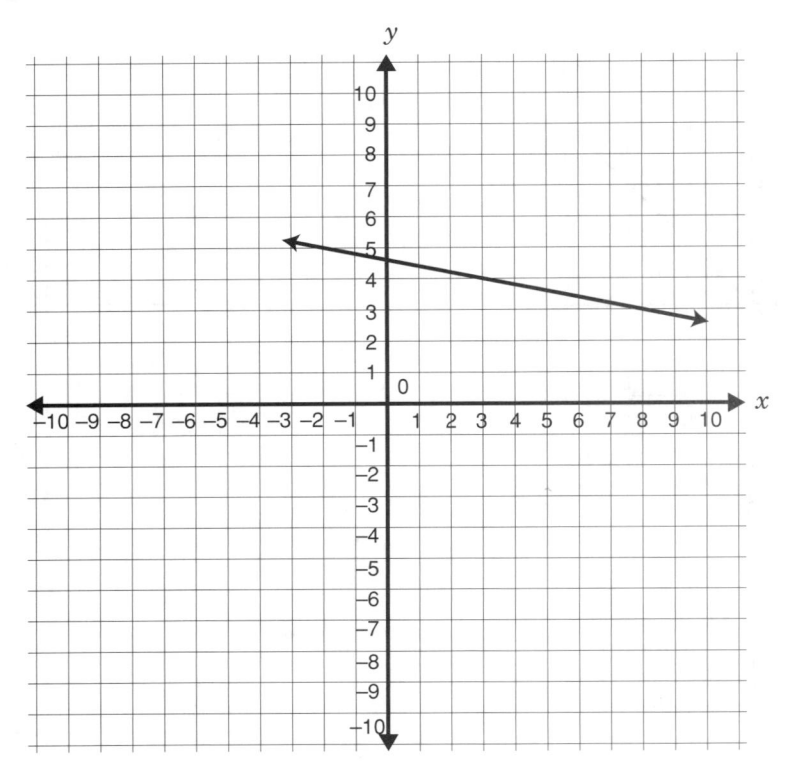

A $-\dfrac{1}{2}$

B $\dfrac{1}{5}$

C $-\dfrac{1}{5}$

D $\dfrac{1}{2}$

30. Juan researched how many students were in his school over the past five years. The results are listed in the chart below.

Class	Year 1	Year 2	Year 3	Year 4	Average Number of Students
Freshman	294	312	301	345	313
Sophomore	445	423	430	428	432
Junior	324	318	312	310	316
Senior	502	495	489	480	492

Based on the information in the table, which of the following statements can be verified?

F Only one class had more than 400 students from Year 1 to Year 4.

G The junior class had the fewest number of students enrolled from Year 1 to Year 4.

H The senior class is double the size of the freshman class.

J The senior class has the greatest average number of students.

31. The net shown below can be folded to represent which figure?

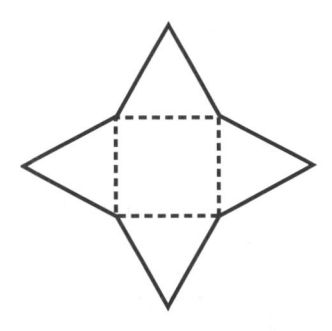

A Cylinder

B Rectangular prism

C Cone

D Pyramid

32. According to the data shown below, which would be the best prediction of the number of widgets produced at the Widget Factory for the year 2012?

Widgets produced at Widget Factory

Year	Number of Widgets
1985	1,321
1990	1,672
1995	2,023
2000	2,374
2005	2,725

F 2,904

G 3,216

H 3,427

J 3,176

GO ON

33. Parallelogram RSTU is shown on the grid below.

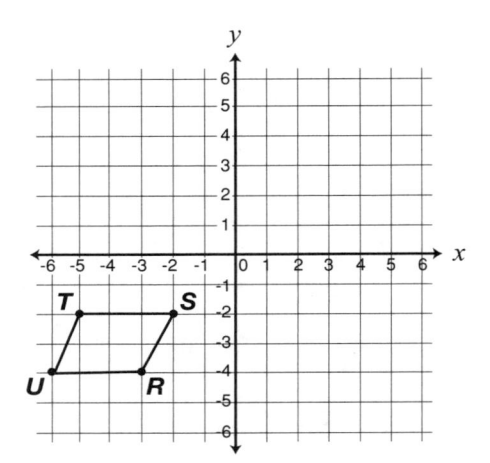

If RSTU is reflected across the line $y = -x$ and then translated 2 units down to become parallelogram R'S'T'U', what will be the coordinates of T'?

A (5, –2)

B (5, 2)

C (3, –5)

D (2, 3)

34. What is the simplified form of $\dfrac{(3a^7b^2)(6ab^4)}{2b^3}$?

F $\dfrac{18a^6}{b^5}$

G $\dfrac{18a^8}{2b^3}$

H $9a^8b^3$

J $\dfrac{9ab^6}{b^3}$

35. The graph of a linear function is shown below.

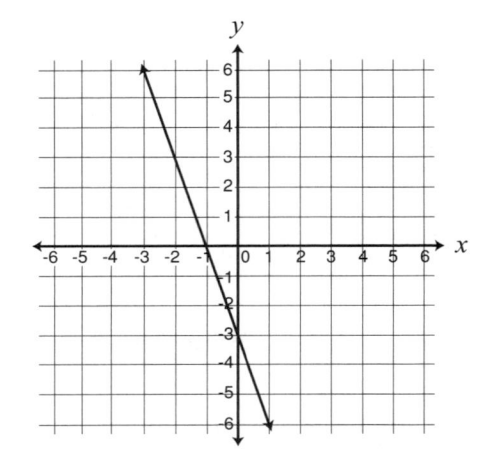

If the line is translated 4 units up, which equation best describes the new line?

A $y = -3x - 3$

B $y = -\left(\dfrac{1}{3}\right)x + 1$

C $y = -3x + 1$

D $y = -\left(\dfrac{1}{3}\right)x - 3$

36. Look at the square shown below.

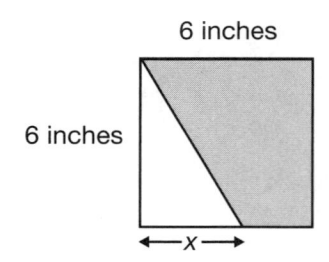

Which expression represents the area of the shaded portion of the 6-inch square?

F $36 + 6x$

G $36 - 3x$

H $12 - 2x$

J $12 + 4x$

37. Line segment $\overline{AB}$ has midpoint M. The coordinates of A are (−4.2, 6.3) and coordinates of point M are (2.4, 4.1). What are the coordinates of point B?

   A (8.1, 6.0)
   B (6.2, 8.4)
   C (3.4, 5.6)
   D (9.0, 1.9)

38. Ramon has $60 to buy several new movie and music DVDs that were recently released. Ramon received an additional $80 for his birthday. If movies cost $20 and music DVDs cost $15, which equation can be used to determine how many movies, y, he can buy if he buys x number of DVDs? Assume Ramon uses all of his money plus birthday money.

   F $y = 9\frac{1}{3} - \left(\frac{4}{3}\right)x$
   G $y = 7 - \frac{3}{4}x$
   H $y = 3 - \frac{3}{4}x$
   J $y = 140 - x$

39. Triangle PRZ has the following vertices: P (−4, 6), R (−6, 3), and Z (−2, 0). Find the coordinates of point M, the endpoint of median $\overline{PM}$, the line drawn from P to the midpoint of line segment $\overline{RZ}$.

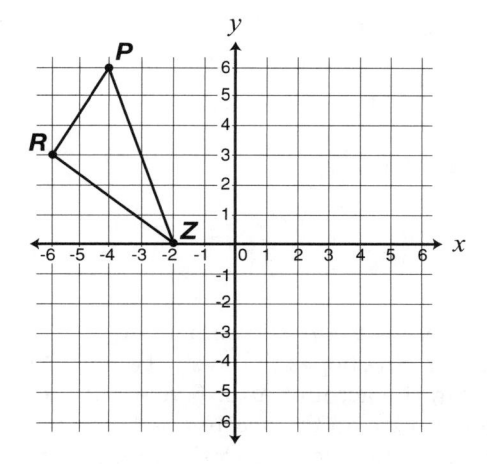

   A (−1.5, 4)
   B (−3, 1.5)
   C (−4, 1.5)
   D (−4, 2.5)

40. What is the length of the line segment with endpoints (7, −6) and (4, −2)?

   Record your answer and fill in the bubbles. Be sure to use the correct place value.

41. Which equation will produce the narrowest parabola when graphed?

A $y = \frac{1}{3}x^2 - 1$

B $y = 2x^2 - 1$

C $y = \frac{1}{2}x^2 + 3$

D $y = -3x^2 + 3$

42. Haley plans to invest some money to save for a new car. She hopes to have between $850 and $900 in the account after 6 years. If the simple interest rate on the account is 4% annually, which of the following initial investments would meet, but not exceed, Haley's goal?

F $650

G $700

H $750

J $800

43. Jane ran a series of tests and recorded her information in the table shown below.

x	y
1	5
2	11
3	21
4	35

Which equation best describes these results?

A $y = 3x + 2$

B $y = 4x - x$

C $y = 2x^2 + 3$

D $y = x^2 - 3x$

44. Wendy is looking out from a point at the top of a ship 200 feet above the water. When Wendy looks down at an angle of depression of 30°, as shown in the figure below, she can see the dock. To the nearest foot, how far is the dock from the base of the ship?

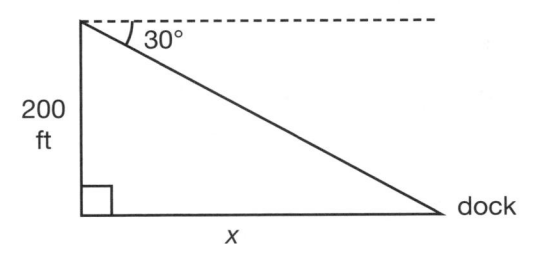

F 283 ft

G 346 ft

H 401 ft

J 1,095 ft

45. Which points best represent the roots of the quadratic equation graphed below?

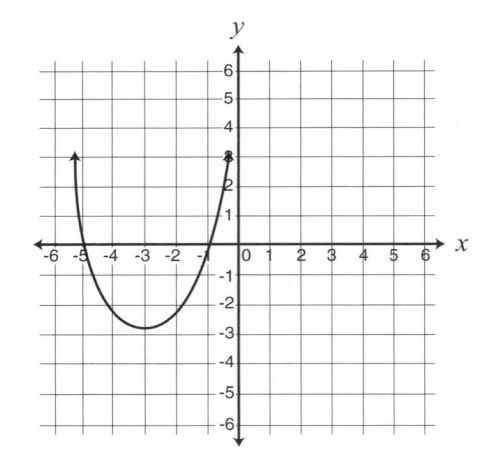

A (−4, −2) and (−2, −2)

B (−5, 0) and (−1, 0)

C $\left(-3, -2\frac{1}{2}\right)$ and (0, 3)

D There are no solutions

46. Movie tickets cost $8.00 for adults and $5.00 for children. A group of adults and children spent a total of $68.00 on movie tickets. If there were a total of 10 people in the group, approximately what percent of the total amount spent was used to buy the children's tickets?

   F 20 percent
   G 25 percent
   H 30 percent
   J 35 percent

47. The graph of the equation $y = -2x - 2$ is given below. Graph $y = \dfrac{1}{2}x + 3$ on the grid.

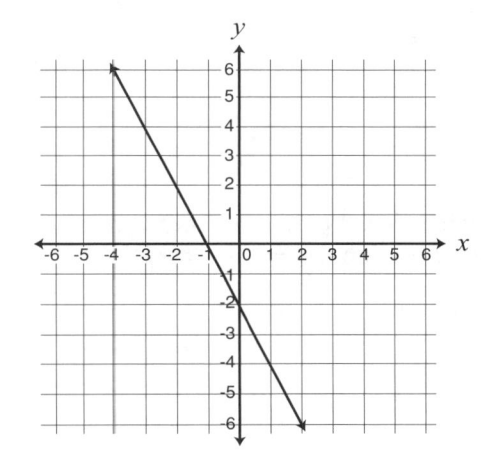

   What is the solution to this system of equations?

   A (–2, 2)
   B (0, 3)
   C (–1, 1)
   D No solution

48. Which set of ordered pairs forms the vertices of a right triangle?

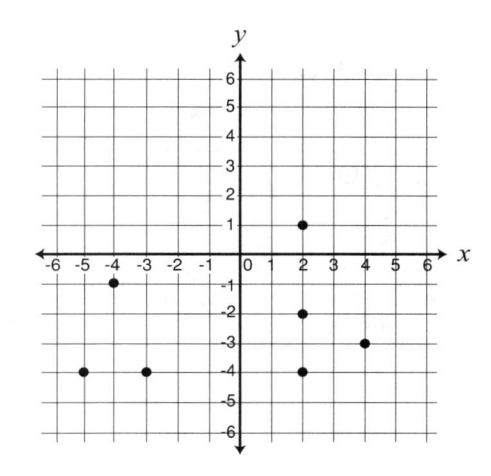

   F (2, –4), (–3, –4), (2, 1)
   G (–5, –4), (–4, –1), (–3, –4)
   H (2, –2), (2, –4), (4, –3)
   J (2, 1), (2, –4), (–4, –1)

49. Find the x- and y-intercepts of $8x - 3y = 24$.

   A x-intercept: (0, –8)
      y-intercept: (3, 0)
   B x-intercept: (–8, 0)
      y-intercept: (0, 3)
   C x-intercept: (0, 3)
      y-intercept: (–8, 0)
   D x-intercept: (3, 0)
      y-intercept: (0, –8)

GO ON

50. Point P is 34 cm from the center of circle O which has a radius of 16 cm.

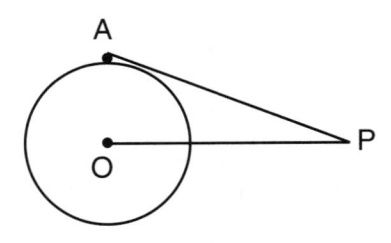

Find the length of the tangent AP drawn to circle O from point P.

F 15 cm
G 17 cm
H 30 cm
J 32 cm

51. Paul went to the grocery store to buy pasta and sauce for a schoolwide spaghetti dinner. He had 72 containers when he left. Pasta costs $1.35 per container. Sauce costs $2.50 per container. Paul spent $140.90 at the grocery store. Which system of linear equations can be used to find p, the number of pasta containers purchased, and s, the number of sauce containers purchased?

A $p + s = 140.90$
   $2.50p + 1.35s = 72$

B $p + s = 140.90$
   $1.35p + 2.50s = 72$

C $p + s = 72$
   $2.50p + 1.35s = 140.90$

D $p + s = 72$
   $1.35p + 2.50s = 140.90$

52. Which equation represents the missing step in the solution?

Step 1: $4(x + 2) - 2 = 8$

Step 2:

Step 3: $4x + 6 = 8$

Step 4: $4x = 2$

Step 5: $x = \dfrac{1}{2}$

F $4x + 6 - 2 = 8$
G $4x + 12 - 6 = 8$
H $4x + 8 - 2 = 8$
J $4x + 6 - 6 = 8$

53. Which ordered pair represents one of the roots of the function $f(x) = x^2 - 4x - 12$?

A (–6, 0)
B (6, 0)
C (1, 0)
D (1/3, 0)

54. How does the graph of $y = -3x^2$ differ from the graph of $y = \left(\dfrac{2}{3}\right)x^2$?

   F The graph of $y = -3x^2$ opens downward and is wider than the graph of $y = \left(\dfrac{2}{3}\right)x^2$.

   G The graph of $y = -3x^2$ opens upward and is wider than the graph of $y = \left(\dfrac{2}{3}\right)x^2$.

   H The graph of $y = -3x^2$ opens downward and is narrower than the graph of $y = \left(\dfrac{2}{3}\right)x^2$.

   J The graph of $y = -3x^2$ opens upward and is narrower than the graph of $y = \left(\dfrac{2}{3}\right)x^2$.

55. There are 50 marbles of different colors in a bag. The percent of marbles of each color is shown below.

Color	Percent
Blue	32
Green	14
Pink	6
Purple	12
Red	5
Yellow	31

If you draw two marbles from the bag without replacing the first marble drawn, which of the following is the approximate probability of drawing two blue marbles?

   A 4.5%

   B 6.8%

   C 7.9%

   D 9.8%

56. If y is directly proportional to x and $y = 8$ when $x = 15$, what is the value of x when $y = -4$?

   F $-7\dfrac{1}{2}$

   G -30

   H $-2\dfrac{1}{3}$

   J $-4\dfrac{3}{4}$

57. Which of the following relationships could *not* be used to determine the length of line segment AR as shown below?

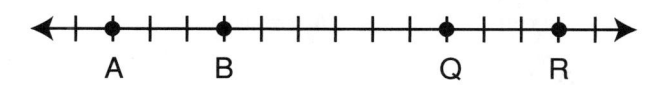

   A $\overline{AB} + \overline{BQ} + \overline{QR} = \overline{AR}$

   B $\overline{AQ} + \overline{BR} = \overline{AR}$

   C $\overline{AQ} + \overline{QR} = \overline{AR}$

   D $\overline{AB} + \overline{BR} = \overline{AR}$

58. Joe's dad decided to pay Joe to do yard work. His dad will pay $7 for every hour of work. Joe buys a rake and a set of bags for $15. In figuring out how much money Joe makes, what would be the dependent quantity?

   F The pay Joe's dad gives each hour

   G The cost of a rake and the set of bags

   H The number of hours Joe works

   J The amount of money Joe makes

GO ON

59. The graph of a system of linear equations is shown below.

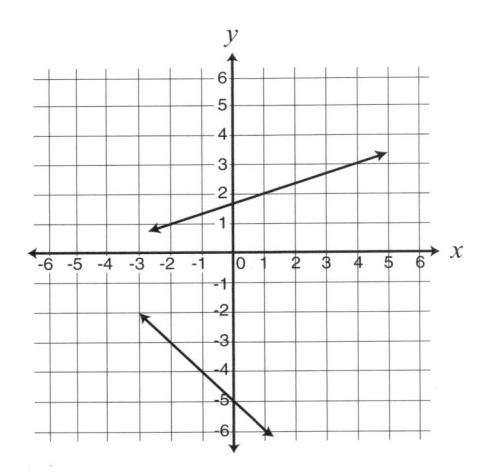

Which of the following is the solution to this system of linear equations?

A (−5, 0)

B (−4, 0)

C (−2, 2)

D (−6, 3)

60. The cost of a trip to the city, c, plus the cost of entertainment, can be represented by the inequality $61d + 35 < c < 67d + 50$, where d represents the number of days the visitor stays in the city. If the person stays 4 days, which of the following is a fair amount to spend during the visit?

F $140.00

G $262.50

H $297.32

J $353.89

BE SURE YOU HAVE RECORDED ALL OF YOUR ANSWERS ON THE ANSWER DOCUMENT

Practice Test 2

1. Ⓐ Ⓑ Ⓒ Ⓓ
2. Ⓕ Ⓖ Ⓗ Ⓙ
3. Ⓐ Ⓑ Ⓒ Ⓓ
4. [grid-in answer box]
5. [grid-in answer box]
6. Ⓕ Ⓖ Ⓗ Ⓙ
7. Ⓐ Ⓑ Ⓒ Ⓓ
8. Ⓕ Ⓖ Ⓗ Ⓙ
9. Ⓐ Ⓑ Ⓒ Ⓓ
10. [grid-in answer box]
11. Ⓐ Ⓑ Ⓒ Ⓓ
12. [grid-in answer box]
13. Ⓐ Ⓑ Ⓒ Ⓓ
14. Ⓕ Ⓖ Ⓗ Ⓙ
15. [grid-in answer box]

16. [grid-in answer]

17. [grid-in answer]

18. Ⓕ Ⓖ Ⓗ Ⓙ
19. Ⓐ Ⓑ Ⓒ Ⓓ
20. Ⓕ Ⓖ Ⓗ Ⓙ
21. Ⓐ Ⓑ Ⓒ Ⓓ
22. Ⓕ Ⓖ Ⓗ Ⓙ
23. Ⓐ Ⓑ Ⓒ Ⓓ
24. Ⓕ Ⓖ Ⓗ Ⓙ

25. [grid-in answer]

26. Ⓕ Ⓖ Ⓗ Ⓙ
27. Ⓐ Ⓑ Ⓒ Ⓓ
28. Ⓕ Ⓖ Ⓗ Ⓙ
29. Ⓐ Ⓑ Ⓒ Ⓓ
30. Ⓕ Ⓖ Ⓗ Ⓙ
31. Ⓐ Ⓑ Ⓒ Ⓓ
32. Ⓕ Ⓖ Ⓗ Ⓙ
33. Ⓐ Ⓑ Ⓒ Ⓓ
34. Ⓕ Ⓖ Ⓗ Ⓙ
35. Ⓐ Ⓑ Ⓒ Ⓓ
36. Ⓕ Ⓖ Ⓗ Ⓙ
37. Ⓐ Ⓑ Ⓒ Ⓓ
38. Ⓕ Ⓖ Ⓗ Ⓙ
39. Ⓐ Ⓑ Ⓒ Ⓓ

40. [bubble grid for numeric answer]

41. Ⓐ Ⓑ Ⓒ Ⓓ
42. Ⓕ Ⓖ Ⓗ Ⓙ
43. Ⓐ Ⓑ Ⓒ Ⓓ
44. Ⓕ Ⓖ Ⓗ Ⓙ
45. Ⓐ Ⓑ Ⓒ Ⓓ
46. Ⓕ Ⓖ Ⓗ Ⓙ
47. Ⓐ Ⓑ Ⓒ Ⓓ

48. Ⓕ Ⓖ Ⓗ Ⓙ
49. Ⓐ Ⓑ Ⓒ Ⓓ
50. Ⓕ Ⓖ Ⓗ Ⓙ
51. Ⓐ Ⓑ Ⓒ Ⓓ
52. Ⓕ Ⓖ Ⓗ Ⓙ
53. Ⓐ Ⓑ Ⓒ Ⓓ
54. Ⓕ Ⓖ Ⓗ Ⓙ
55. Ⓐ Ⓑ Ⓒ Ⓓ
56. Ⓕ Ⓖ Ⓗ Ⓙ
57. Ⓐ Ⓑ Ⓒ Ⓓ
58. Ⓕ Ⓖ Ⓗ Ⓙ
59. Ⓐ Ⓑ Ⓒ Ⓓ
60. Ⓕ Ⓖ Ⓗ Ⓙ

BE SURE YOU HAVE RECORDED ALL OF YOUR ANSWERS ON THE ANSWER DOCUMENT

STOP

Practice Test 2 Answer Explanations

1. B   08   G(e)(1)(A)

You don't need to find the length of the hypotenuse to solve this problem. Use the formula $A = \frac{1}{2}bh$:

$A = \frac{1}{2}(10 \times 12) = 60 \text{ ft}^2$

2. G   02   A(b)(3)(B)

Frankie had 30 baseball cards at the beginning of the week and he gave x away. So, you know the expression will begin with $30 - x$. Then Frankie's mother gives him y more baseball cards, so the entire expression should be $30 - x + y$.

3. D   08   G(e)(1)(A)

To solve this problem, you need to find the area of a circle. To do this, use the formula $A = \pi r^2$. The radius is 12, and 12 squared is 144. The number 144 multiplied by 3.14 is about 452.

4. 4 m   08   G(e)(1)(D)

To solve this problem, you need to substitute values into the formula for the surface area of a sphere:

$S = 4\pi r^2$

$201 \approx 4(3.14)r^2$

$201 \approx 12.56 r^2$

$\frac{201}{12.56} \approx r^2$

$16 \approx r^2$

$r \approx 4 \text{ m}$

5.   12 ft   08   G(e)(1)(C)

$a^2 + b^2 = c^2$

$9^2 + b^2 = 15^2$

$b^2 = 15^2 - 9^2$

$b^2 = 225 - 81$

$b^2 = 144$

b = 12 ft

6.   H   08   G(e)(1)(D)
A cube has the same, length, width, and height, so you would substitute 4 inches into the formula $V = lwh$.

7.   C   05   A(d)(3)(A)
If the initial population sample was 100 butterflies and it doubled every 5 days, it would double twice in 10 days, to 200 on day 5 and then to 400 on day 10.

8.   F   07   G(e)(2)(D)
The figure is a square pyramid. If you count the faces, edges, and vertices, you'll see that the correct answer is answer choice F.

9.   B   02   A(b)(4)(A)
The expression $4(x + 2y) + 2(3x - y) - (x + y)$ can be simplified like this: $4x + 8y + 6x - 2y - x - y$. Then combine like terms to get $9x + 5y$.

10. 450   09   (8.3)(B)

Let the amount that Ray earns equal x and the amount that Cory earns equal $1.5x$.

$x + 1.5x = \$750$

$2.5x = \$750$

$x = \$300$

Ray earns $300, so

$750 – 300 = \$450$

Cory earns $450.

11. A   08   G(f)(1)(A)

This is a graph resembling that of an equation of the form $y = a \cdot b^x$, which is exponential (a and b are constants).

12. $127.39   10   (8.14)(A)

To solve this problem, you first need to determine the 35% discount. Multiply .35 by $195.99, the price of the bike before the discount. Then deduct the result from $195.99 to find how much Peter paid after the refund.

13. B   10   (8.16)(B)

You can use your calculator to quickly determine this answer: 225 is 15^2.

14. H   02   A(b)(4)(B)

$$\frac{3y}{y^3} = \frac{3 \cdot y}{y \cdot y^2} = \frac{3}{y^2}$$

15. $67.99   10   (8.14)(B)

First, determine how much the sneakers were discounted: $84.99 \times .20 = \$17.00$. Then deduct this amount from the original cost of the sneakers.

16. 17.68   09   (8.3)(B)

To find the number of people employed in manufacturing jobs, multiply 136,000,000 by 13%, or .13. The answer comes out to 17,680,000, or 17.68 million.

17. $416   02   A(b)(3)(A)
Melanie worked 40 hours for which she earned $8 per hour. She earned $320 for the 40 hours. Then she worked 8 hours of overtime, for which she was paid $12 per hour so she earned $96 in overtime. If you add $320 + $96, the answer is $416.

18. J   01   A(b)(1)(D)
You can substitute values to find the correct graph. Remember that when only the < sign is present, the line should be dashed. Shading should be below the line.

19. B   02   A(b)(4)(A)
Divide both sides of the inequality by 6: $\frac{6y}{6} < \frac{42}{6}$.

20. H   09   (8.3)(B)
If you multiply 570 by 4, the answer is 2,280.

21. C   09   (8.3)(B)
To find the discount, multiply $75 by .20. The discount is $15.

22. J   06   G(c)(1)(A)
These angles are supplementary; they add up to 180 degrees. To find the answer subtract the value of the given angle, 45 degrees, from 180 degrees.

23. D   09   (8.3)(B)
Each week, Madeline earns $96. Over 10 weeks, that equals $960.

24. G   09   (8.12)(A)
If you subtract the age of the youngest person, y, from the age of the oldest person, you'll get the range, or 62 − y = 51. Therefore, the youngest person is 11.

25. 40   01   A(b)(1)(D)
If $y = \left(\frac{1}{4}\right)x$, and $y = 10$, then $10 = \left(\frac{1}{4}\right)x$, and $x = 40$ inches.

26. H   06   G(c)(1)
If the letter R is first reflected and then slid, it would look like the transformation in answer choice H.

27. D   03   A(c)(2)
Parallel lines have the same slope.

28. F   03   A(c)(1)(A)
A linear function graphs as a non-vertical straight line.

29. C   01   A(b)(1)(C)
Use the formula, $\dfrac{y_2 - y_1}{x_2 - x_1}$ with points (–2, 5) and (8, 3) to find the slope $\dfrac{3-5}{8-(-2)} = \dfrac{-2}{10} = \dfrac{-1}{5}$.

30. J   09   (8.13)(B)
The only answer choice that is supported by the information in the table is J: the senior class has the greatest average number of students.

31. D   07   G(d)(1)(B)
Folding the triangles up so that they meet in the middle will show that the net is a pyramid.

32. G   01   A(b)(1)(B)
The number changes by 351 every 5 years. To find the change in 7 years (from 2005 to 2012) at the same rate, set up the proportion $\dfrac{351}{5} = \dfrac{y}{7}$. Cross-multiply to get $5y = 2457$, and $y = 491.4$. This is the change from 2005 to 2012, so add it to the value for 2005 to get the prediction for 2012: $2{,}725 + 491.4 = 3{,}216.4$.

33. D   06   G(e)(3)(A)
When you reflect a point across the diagonal line $y = -x$, the x-coordinate and the y-coordinate change places and signs. Point T has coordinates (–5, –2). When you reflect T across the line $y = -x$, you change the places of the coordinates and the signs, making the coordinates (2, 5). Now, you need to translate the coordinates 2 units down to make the parallelogram R'S'T'U'. The coordinates for T' would then have a y value 2 units less, and would be (2, 3).

34. H 05 A(d)(3)(A)

Use the multiplication rule for exponents to simplify this expression. The multiplication rule tells you to multiply coefficients and add exponents of similar variables: $\frac{(3)(6)(a^{7+1})(b^{2+4})}{2b^3}$. Now divide the similar variables that appear in the numerator and the denominator by subtracting their exponents (division rule for exponents): $\frac{18}{2} \times a^8 \times b(b^{-3}) = 9a^8b^3$.

35. C 03 A(c)(2)(C)

The slope is found by calculating the change in y divided by the change in x, so the slope $= -3$, and the new line will have the same slope because it is only being translated upward. The line currently intersects with the y-axis at $y = -3$. Raising the intersection 4 units would change the y-intercept to $y = 1$. So by the slope-intercept formula $y = mx + b$, $y = -3x + 1$.

36. G 06 G(b)(4)(A)

To solve this problem, first find the area of the square and then subtract the unshaded triangle. For this square, the expression used to calculate the area ($A = s^2$) would be $A = 6^2 = 36$. Next, to find the area of the triangle, notice that the height of the triangle is one side of the square, or 6, and the triangle's base is represented by x. For this triangle, the area $A = \frac{1}{2}bh$ would be $A = \frac{1}{2} \times 6x = 3x$. So the area of the shaded portion is $36 - 3x$.

37. D 07 G(d)(2)(C)

Use the midpoint formula to figure out the answer. First substitute the x-coordinate values of points M and A into the formula.

$2.4 = \frac{(-4.2 + x_2)}{2}$

$4.8 + 4.2 = x_2$

$x_2 = 9.0$

Next, substitute the y-coordinate values.

$$4.1 = \frac{(6.3 + y_2)}{2}$$

$$8.2 = (6.3 + y_2)$$

$$8.2 - 6.3 = y_2$$

$$y_2 = 1.9$$

The coordinates of point B are therefore (9.0, 1.9)

38. G   04   A(c)(3)(A)

If Ramon uses all of his money, then the total cost of movie and music DVDs equals $60 + $80 = $140. If x = the number of music DVDs Ramon buys, then $15x$ is the amount of money he spends on music. Then $20y$ is the amount of money Ramon spends on movies. The initial formula looks like this:

$$15x + 20y = 60 + 80$$

That equation is simplified to $15x + 20y = 140$. Subtract $15x$ from each side to get:

$$20y = 140 - 15x$$

Then divide each side by 20 to get:

$$y = \frac{140}{20} - \left(\frac{15}{20}\right)x$$

which simplifies to:

$$y = 7 - \frac{3}{4}x$$

Answer choice G is the correct answer.

39. C   07   G(d)(2)(C)

Use the midpoint formula to figure out the answer.

$$M = \left(\frac{(x_1 + x_2)}{2}, \frac{(y_1 + y_2)}{2}\right)$$

The coordinates for line $\overline{RZ}$ are $(x_1, y_1) = (-6, 3)$ and $(x_2, y_2) = (-2, 0)$. The coordinates for M are therefore $(-4, 1.5)$.

40. d = 5   07   G(d)(2)(C)
To solve this problem you need to use the distance formula.

$$d = \sqrt{(x_2 - x_1)^2 + (y_2 - y_1)^2}$$

$$d = \sqrt{(4-7)^2 + (-2-(-6))^2}$$

$$d = \sqrt{(-3)^2 + (4)^2}$$

$$d = \sqrt{9 + 16}$$

$$d = \sqrt{25}$$

$$d = 5$$

41. D   05   A(d)(1)(B)
The width of a parabola is determined by the coefficient of the x^2 variable. The lowest absolute value of this coefficient will have the widest arc and the highest absolute value of this coefficient will have the narrowest arc. Answer choices B and C are both wrong because the coefficients are less than 1. For answer choice B, the absolute value of 2 is 2. For answer choice D, the absolute value of –3 is 3. Because 3 is the highest number, it has the narrowest parabola.

42. G   10   (8.14)(C)
To solve this problem use the formula $I = prt$, where I is the amount of interest earned, p is the principal or amount initially invested, r is the interest rate (4%, or 0.04), and t is the time in years that the money is invested (6 years). The only initial investment answer choice that produces a balance between $850 and $900 after 6 years is $700, which produces a balance of $868.

43. C   01   A(b)(1)(B)
To solve this problem, insert the number for x with the number for y to see which formula is correct. In answer choice A, when $x = 1$, $y = 5$, but when $x = 2$, $y = 8$, so it cannot be correct because the table says $y = 11$ when $x = 2$. For answer choices B and D, y does not equal 5 when $x = 1$, so they cannot be right. The formula in answer choice C is correct because all of the values of x and y in that formula match the values of x and y in the table.

44. G   06   G(c)(1)(C)

The triangle in the diagram is a 30°– 60°– 90° triangle. The shorter leg of the triangle is 200 feet. The length of the longer leg is the distance from the dock to the base of the ship, or x. With this type of triangle, the length of the longer leg is equal to the length of the shorter leg times $\sqrt{3}$. The dock is $200 \times \sqrt{3}$ from the base of the ship. Use your calculator find $\sqrt{3}$ and multiply this number by 200. The answer is 346.41. This means that the dock is approximately 346 feet from the base of the ship.

45. B   05   A(d)(2)(B)

The roots of a graphed quadratic equation are the points where the parabola crosses the x-axis. From the graph, these points are $(-5, 0)$ and $(-1, 0)$, or answer choice B.

46. H   10   (8.14)(C)

To solve this problem, you need to write two equations. First, you need to figure out the number of adults, a, and children, c, in the group. This can be represented by the equation

$$a + c = 10$$

The next equation will have to do with prices. The number of adults times the adult ticket price plus the number of children times the child ticket price equals the total cost. This expression can be represented with the equation $8.00a + 5.00c = 68.00$, or

$$8a + 5c = 68$$

To solve the system of equations, use the first equation to substitute for a: $a = 10 - c$ in the second equation:

$$8(10 - c) + 5c = 68$$

$$80 - 8c + 5c = 68, \text{ or}$$

$$3c = 12, \text{ which gives } c = 4.$$

So there were 4 children in the group. Now, find the cost for the children's tickets.

4 children × $5/ticket = $20

To find the percent of the total amount spent to buy the children's tickets, solve the proportion $\frac{20}{68} = \frac{x}{100}$.

By cross-multiplication, $68x = 2000$, or x is approximately 30%.

47. A   04   A(c)(4)(B)

Sketch the new equation $\left(y = \frac{1}{2}x + 3\right)$ on the graph first. The y-intercept is +3 so make a point at the place (0, 3) on the graph. The slope is $\frac{1}{2}$ so the line will rise (y) one unit and run (x) two units. So one point can be at location (2, 4) and another point can be at location (–2, 2). Draw a straight line to connect the three dots and extend it across the entire graph. Look at the location the two lines intersect, (-2, 2). That is the solution to the set of equations.

48. F   07   G(d)(2)(A)

The two sides of the triangle must be perpendicular to form a right triangle. In answer choice A, the line segment with endpoints (2, –4) and (2, 1) is vertical, and the line segment with the endpoints (2, –4) and (–3, –4) is horizontal, so they are perpendicular.

49. D   03   A(c)(2)(E)

To find the x- and y-intercepts, substitute 0 for the other variable. That is, replace y with 0 to find the x-intercept, and replace x with 0 to find the y-intercept.

$8x - 3y = 24$

$8x - 3(0) = 24$

$8x - 0 = 24$

$8x = 24$

$x = 3$

The x-intercept is 3. The line crosses the x-axis at (3, 0).

$8x - 3y = 24$

$8(0) - 3y = 24$

$0 - 3y = 24$

$-3y = 24$

$y = -8$

The y-intercept is –8. The line crosses the y-axis at (0, –8)

Answer choice D is the only answer that lists both the x-intercept and the y-intercept correctly.

Practice Test 2 Answer Explanations

50. H   06   G(c)(1)(C)
Triangle PAO is a right triangle. Since (8, 15, 17) is a Pythagorean triple, its double is also a Pythagorean triple: (16, 30, 34). This means line $\overline{AP}$ is 30 centimeters long.

51. D   04   A(c)(4)(A)
Answer choices A and B are both incorrect. The values p and s should add to 72, the number of containers Paul bought at the grocery. The total cost ($140.90) belongs in the equation that includes the appropriate cost per item (1.35 for p and 2.50 for s). Answer choice D is the correct answer. The total number of containers ($p + s$) is 72. The total cost is $140.90, and the cost for pasta, 1.35p, and sauce, 2.50s, are each correct.

52. H   10   (8.15)(A)
Step 1 has an expression in parentheses. Simplify the expression by multiplying to remove the parentheses first: $4x + 8 - 2 = 8$.

53. B   05   A(d)(2)(B)
To solve this question, replace the x value with the value in the ordered pair, and find which equation makes $f(x) = 0$. For answer choice A, the equation equals 48, so answer A is not correct. For answer choice B, the equation equals 0, which appears to be correct, but you should check each of the other options. The equation equals –15 for answer choice C and it equals $\dfrac{-119}{9}$ in answer choice D. So answer choice B is correct.

54. H   05   A(d)(1)(B)
The graph of $y = -3x^2$ opens downward because the coefficient of x^2 is negative, so answer choices G and J are both incorrect. The absolute value of -3 is 3. The absolute value of $\dfrac{2}{3}$ is $\dfrac{2}{3}$. Because 3 is greater than $\dfrac{2}{3}$, the graph of $y = -3x^2$ is more narrow, than the graph of $y = \left(\dfrac{2}{3}\right)x^2$.

55. D   10 (8.14)(B)

This problem involves two dependent events. The favorable outcome is drawing a blue marble. Of the 50 marbles in the bag, 32% are blue, so the probability of the first marble being blue is .32, or

$$P(\text{blue first}) = \frac{16}{50} = 0.32$$

To find out the number of blue marbles, multiply the percent by the number of total marbles, $0.32 \times 50 = 16$ blue marbles for the first pick.

Now there are only 49 marbles in the bag, and assuming the first draw was blue, there are 15 blue marbles still in the bag, or

$$P(\text{blue second}) = \frac{15}{49} = 0.3061$$

To find the probability of a compound event, multiply the probabilities.

$0.32 \times 0.3061 = 0.097952$

This means that the probability of drawing two blue marbles from the bag is approximately 9.8%.

56. F   03 A(c)(2)(G)

If x and y are directly proportional, the ratios between the variables are equal, or

$$\frac{x_1}{y_1} = \frac{x_2}{y_2}$$

Substitute values for the variables, or

$$\frac{15}{8} = \frac{x}{-4}$$

Cross-multiply to get

$15 \times -4 = 8x$

$-60 = 8x$, or

$$x = -7\frac{1}{2}$$

57. B   07 G(d)(2)(A)

The only formula that could not be used to determine the length of line segment $\overline{AR}$ would be $\overline{AQ} + \overline{BR} = \overline{AR}$. $\overline{AQ} + \overline{BR}$ would be longer than the line segment.

58. J   01   A(b)(1)(A)

Answer choices F and G are both wrong because both answers are constants, not variables. Answer choice H is wrong because the number of hours that Joe works is the independent quantity—he can work for as long as he wants. Answer choice J is the dependent quantity because it depends on how long Joe works.

59. A   04   A(c)(4)(B)

Look for the point where the two lines intersect to find the correct answer for this question. Take a ruler or book edge to make sure you keep the lines straight. The two lines would intersect at the point (–5, 0) so answer choice A is the correct answer. You can check this by substituting (–5, 0) into both equations to see that it is a solution to both. The two equations are $y = \frac{1}{3}x + \frac{5}{3}$ and $y = -x - 5$.

60. H   04   A(c)(3)(C)

To solve this question, replace the number of days, 4, for the value d in the inequality. The result is $279.00 < c < $318. Of the four choices, only answer choice C, $297.32, falls within that range.

Index

A
Absolute functions, 89
Absolute value, 14
Accommodations, 2
Acute angles, 119
Addition
    associative property, 38–39
    commutative property, 37–38
Adjacent angles, 120
Algebraic equations, 42–43
Algebraic expressions, 40–41
Angles, 119–121, 126
Area, 156–158
Associative property, 38–39

B
Bar graphs, 181–185

C
Center, of circle, 126
Center of rotation, 137
Central angles, 126
Circle graphs, 182–183
Circles
    area, 157
    central angles, 126
    circumference, 152, 153
    description, 113
Circumference, 152, 153
Coefficients, 40
Commutative property, 37–38
Complementary angles, 120
Cones, 135
Congruent figures, 114
Conversions
    decimal into percent, 18
    fraction into decimal, 17
    percent into decimal, 18
    percent into fraction, 18–19
Coordinate planes, 136–137
Coordinates, 60
Cube of number, 19
Cubic functions, 91
Cylinders
    description/properties, 135, 158
    surface area, 160
    volume, 159

D
Decimals, 16–18
Denominators, 14
Dependent variables, 58, 60
Disabled students, 2
Discounts, 188–189
Distributive property, 40
Domain, 92–93

E
Edges, 134
Endpoints, 118
English language learners (ELL), 2
Equations, 42–43, 66–69
Equilateral triangles, 123, 124
Equivalent fractions, 14–15
Equivalent numbers, 14
Exponential functions, 90
Exponents, 19–20
Expressions, 40–41, 66–67

F
Faces, 134
Fractions, 14–16, 17
Functions
    coordinates, 60
    description, 59–60
    domain and range, 92–93
    linear, 83–86, 94–95
    parent, 88–92
    quadratic, 87–88
    vertical line test, 60–61

G
Graphs, of data, 181–183
Graphs, of functions
    absolute functions, 89
    cubic functions, 91
    exponential functions, 90
    inequalities, 65–66
    linear functions, 84–86
    quadratic functions, 87–88
    vertical line test, 60–61

H
Hexagons, 113, 152
Hypotenuse, 123, 124–125

I
Independent variables, 58
Inequalities, 44, 65–66
Integers, 13
Interest, 189
Inverse operation, 22, 26
Irrational numbers, 13
Isosceles triangles, 123

L
Linear equations, 67–68
Linear functions, 83–86, 94–95
Line graphs, 181
Lines, 94, 118
Line segments, 118

Index

M
Mean, 176
Measures of central tendency, 170, 176–178
Median, 176
Mixed numbers, 16
Mode, 177
Money problems, 187–189
Multiplication
    associative property, 38–39
    commutative property, 37–38
    of fractions, 15
    of radicals, 22

N
Nets, 142
Numbers, types of, 13
Numerators, 14

O
Obtuse angles, 119
Octagons, 113
Opposites, 13
Order of operations, 25, 36–37
Origin, 136

P
Parabolas, 87–88
Parallel lines, 118, 121
Parallelograms, 112, 157
Parent functions, 88–92
Patterns, 45
Pentagons, 113
Percents, 18–19
Perimeter, 152–153
Perpendicular lines, 118
Pi (π), 153
Pie charts or graphs, 182–183
Plane figures, 112–115, 142
Point of intersection, 118
Powers, 19–20
Principal, 189
Prisms
    description/properties, 135, 158
    surface area, 160
    volume, 159
Probability, 170–172
Problem solving, 201–202
Properties, 37–40
Pyramids, 135, 159
Pythagorean theorem, 124–125

Q
Quadratic functions, 87–88
Quadrilaterals, 112

R
Radicals, 22
Radicands, 22
Radius, 126, 153
Range, 92–93, 177–178
Rational numbers, 13
Rays, 118
Real numbers, 13
Rectangles
    area, 156
    congruent, 114
    description, 112
    similar, 115
Rectangular prisms, 158, 159, 160
Rectangular solids, 135
Reducing fractions, 18–19
Reflection, 138
Reflex angles, 119
Repeating decimals, 18
Rhombuses, 112
Right angles, 119
Right circular cones, 135
Right circular cylinders
    description/properties, 135, 158
    surface area, 160
    volume, 159
Right triangles, 123
Rise over run formula, 94
Rotation, 137

S
Sale prices, 188–189
Scalene triangles, 123
Scatterplots, 98–99
Scientific notation, 20–22
Similar figures, 114–115
Slope, 94
Special education students, 2
Spheres, 135, 159
Square of number, 19
Square pyramids, 135, 159
Square roots, 22
Squares, 112
Standards, 4–12
Straight angles, 119
Supplementary angles, 120
Surface area, 160

T
Terminating decimals, 18
Texas Assessment of Knowledge and Skills (TAKS)
    about, 1–2
    accommodations, 2
    overview, 3
    standards, 4–12
    study tips, 3
    when/where given, 2
Three-dimensional figures
    alternate views, 142–143
    nets, 142
    parts of, 134
    surface area, 160
    types of, 135
    volume, 158–160

Transformations, 137–139
Translation, 138
Transversals, 121
Trapezoids, 113, 156
Triangles
    area, 156
    congruent, 114
    description, 113
    description/properties, 123–124
    Pythagorean theorem, 124–125
    similar, 114
    types of, 123
Triangular prisms, 135

V

Variables
    definition, 37
    expressions, 40
    functions, 58–59, 59–61

Venn diagrams, 183–184
Vertex (vertices)
    definition, 119
    parabolas, 87
    three-dimensional figures, 134
    triangles, 123
Vertical angles, 120
Vertical line test, 60
Volume, 158–160

X

x-axis, 136
x-coordinates, 92
x-intercepts, 93

Y

y-axis, 136
y-coordinates, 92
y-intercepts, 93, 94–95